W9-CRB-454

Uniscience Series on Fine Particle Science and Technology

Editor

John Keith Beddow, Ph.D.
Professor of Chemicals and Materials Engineering
Division of Materials Engineering
University of Iowa
Iowa City, Iowa

Advanced Particulate Morphology
J. K. Beddow and T. P. Meloy

Separation of Particles From Air and Gases, Volumes I and II
Akira Ogawa

Particle Characterization in Technology
Volume I: Applications and Microanalysis
Volume II: Morphological Analysis
J. K. Beddow, Editor

Separation of Particles From Air and Gases

Volume II

Author

Akira Ogawa, Dr. Eng.
Associate Professor
Department of Mechanical Engineering
College of Engineering
Nihon University
Fukushima, Japan

Uniscience Series on Fine Particle Science and Technology

Editor-in-Chief

John Keith Beddow, Ph.D.
Professor of Chemicals and Materials Engineering
Department of Materials Engineering
University of Iowa
Iowa City, Iowa

CRC Press, Inc.
Boca Raton, Florida

Library of Congress Cataloging in Publication Data

Ogawa, Akira, 1940 —
Separation of particles from air and gases.

(Uniscience series on fine particle science technology)
Includes bibliographical references and index.
1. Dust—Removal—Equipment and supplies. 2. Dust control. 3. Air—Cleaning. 4. Gases—Cleaning.
I. Title. II. Series.
TH7692.035 1984 628.5'3 82-20679
ISBN 0-8493-5787-X (v.1)
ISBN 0-8493-5788-8 (v.2)

This book represents information obtained from authentic and highly regarded sources. Reprinted material is quoted with permission, and sources are indicated. A wide variety of references are listed. Every reasonable effort has been made to give reliable data and information, but the author and the publisher cannot assume responsibility for the validity of all materials or for the consequences of their use.

Direct all inquiries to CRC Press, Inc., 2000 Corporate Blvd., N.W., Boca Raton, Florida, 33431.

International Standard Book Number 0-8493-5787-X (Volume 1)
International Standard Book Number 0-8493-5788-8 (Volume 2)

Library of Congress Card Number 82-20679

Printed in the United States

PREFACE

Dr. John Aitken wrote in his paper (On the formation of small clear spaces in dusty air, *Trans. R. Soc. Eng.*, 32, 1883-1884) the following:

> The dust particles floating in our atmosphere are every day demanding more and more attention. As our knowledge of these unseen particles increases, our interest deepens, and I might almost say gives place to anxiety, when we realize the vast importance these dust particles have on life, whether it be those inorganic ones so small as to be beyond the powers of the microscope, or those larger organic ones which float unseen through our atmosphere, and which, though invisible, are yet the messengers of sickness and of death to many — messengers far more real and certain than poet or painter has ever conceived . . .

Recently, with developments in science and technology, physical and chemical properties of solid particles, dust, and fumes have become clearer in comparison to the knowledge of particle properties of about 50 or 100 years ago. Dust collectors for air pollution and the chemical and mechanical industries must be planned corresponding to the new knowledge of these dust properties.

In order to design new types of dust collectors, it is necessary not only to rely on usual experience and techniques, but also to grasp the fundamental mechanisms and behaviors of the fine solid particles in the turbulent gas flow occurring in ducts or separation chambers. Then, in order to urge the new creative ideas for the design of dust collectors, the author describes the motion of coarse and fine solid particles in turbulent gas flows in detail. Further, from the fluid dynamical point of view, many kinds of the fundamental constructions of dust collectors are shown with many illustrations.

ACKNOWLEDGMENTS

The author wishes to express his gratitude to Prof. Dr. John Keith Beddow of the University of Iowa for his advice, encouragement, and suggestions.

Further, it is a pleasure to acknowledge the contributions of Prof. Dr. I. Tani of Nihon University, Tokyo; Prof. Dr. K. Iinoya of Aichi K. University, Nagoya; Dipl.-Ing. H. Klein and Dipl.-Phys. R. Pieper of Siemens-Schuckertwerken, Erlangen; Prof. Dr. K. Melcher, Dipl.-Ing. J. Komaroff, and Dipl.-Ing. K. Ito of Technisches Zentrum; Robert Bosh, Stuttgart; Dr. M. P. Escudier of Brown Boveri, Baden; Prof. Dr. E. Muschelknautz of Bayer AG, Dormagen; Prof. Dr. H. Brauer of Technische Universität, Berlin; Prof. Dr. M. Bohnet of Technische Universität, Braunschweigh; Prof. Dr. F. Löffler of Technische Universität, Karlsruhe; Dr. J. C. Rotta of DFVLR, Göttingen; Prof. Dr. O. Molerus of Technische Universität, Erlangen-Nürnberg; Prof. Dr. C. Alt of Technische Universität, Stuttgart; Prof. Dr. J. O. Hinze of Delft University, Delft; Prof. Dr. A. Fortier, Prof. Dr. R. Comolet, and Dr. C. P. Chen of the University of Paris (VI); Prof. Dr. J. Abrahamson of Canterbury University, New Zealand; Dr. W. H. Gauvin of Noranda, Québec; Prof. Dr. P. L. Corbella of the University of Barcelona, and Prof. Dr. E. Costa Novella of the University of Madrid, Spain; and Prof. Dr. R. H. Page, of Texas A & M University.

The author should like to express his gratitude to the following authors:

Beddow, J. K., *Particulate Science and Technology,* Chemical Publishing, New York, 1982.

Bethea, R. M., *Air Pollution Control Technology,* D. Van Nostrand, New York, 1978.

Blacktin, S. C., *Dust,* Chapman & Hall, London, 1934.

Brauer, H. and Varma, Y. B. G., *Air Pollution Control Equipment,* Springer-Verlag, Berlin, 1981.

Comolet, R., Dynamiques des fluides réeles, *Tome II,* Masson et Cie, Paris, 1982.

Crawford, M., *Air Pollution Control Theory,* McGraw-Hill, New York, 1976.

Davies, C. N., *Aerosol Science,* Academic Press, New York, 1966.

Dorman, R. G., *Dust Control and Air Cleaning,* Pergamon Press, Oxford, 1974.

Faraday Society, *Disperse Systems in Gases; Dust, Smoke and Fog,* Gurney & Jackson, London, 1936.

Fortier, A., *Mécanique des Suspensions,* Masson et Cie, Paris, 1967.

Iinoya, K., *Dust Collection,* (in Japanese), Nikkan-Kogyo-Shinbun-Sha, Tokyo, 1980.

Green, H. L. and Lane, W. R., *Particulate Clouds, Dusts, Smokes and Mists,* E. & F. N. Spon Ltd., London, 1964.

Kano, T., *Motion of the Solid-Particles,* (in Japanese), Sangyo-Gizitzu Center, Tokyo, 1977.

Ledbetter, J. O., *Air Pollution, Part B, Prevention and Control,* Marcel Dekker, New York, 1974.

Lewis, A., *Clean the Air,* McGraw-Hill, New York, 1965.

Lin, B. Y. H., *Fine Particles,* Academic Press, New York, 1976.

Mallette, F. S., *Problems and Control of Air Pollution,* Van Nostrand, New York, 1955.

Marchello, J. and Kelly, J. J., *Gas Cleaning for Air Quality Control,* Marcel Dekker, 1975.

Meldau, R., *Handbuch der Staubtechnik,* Erster and Zweiter Volumes, VDI-Verlag, Düsseldorf, 1958.

Morikawa, Y., *Two-Phase Flow of Fluid-Solid,* (in Japanese), Nikkan-Kogyo-Shinbun-Sha, Tokyo, 1980.

Ogawa, A., *Cyclone Dust Collectors,* (in Japanese), Earth Publishing, Tokyo, 1980.

Ogawa, A., *Vortex Flow,* (in Japanese), Sankaido Publishing, Tokyo, 1981.

Parker, A., *Industrial Air Pollution Handbook,* McGraw-Hill, London, 1978.

Richardson, E. G., *Aerodynamic Capture of Particles,* Pergamon Press, Oxford, 1960.

Schweitzer, P. A., *Handbook of Separation Techniques for Chemical Engineerings,* McGraw-Hill, New York, 1979.

Štorch, O., *Industrial Separators for Gas Cleaning,* Elsevier, Amsterdam, 1979.

Thring, M. W., *Air Pollution,* Butterworths, London, 1957.

Weber, E. and Broke, W., *Apparate and Verfahren der industriellen Gasvereinigung,* Vol. 1, O. Oldenbourg Verlag, Munich, 1973.

The Institution of Chemical Engineers, *A Users Guide to Dust and Fume Control,* Institute of Chemical Engineering, London, 1981.

Zimon, A. D., *Adhesion of Dust and Powder,* Plenum Press, New York, 1969.

Whytlaw-Groy, R. and Patterson, H. S., *Smoke,* Edward Arnold, London, 1932.

Gordon, G. M. and Pejsahov, I. L., Pyleuablavlivanie i Ochistka Gazov v Tsvetnoi Metallurgii, *Moskva Metallurgiia,* 1977.

Grigorjev, M. A. and Pokrovskij, G. P., *Avtomobiljnye i Traktornye Tsentrifugi,* Moskva, 1961.

Finally, I should like to thank the Managing Editor of CRC Press, Sandy Pearlman and Anita Hetzler, Coordinating Editor, for their suggestions.

These volumes could not have been completed without their kind help.

Akira Ogawa
July 1982, Tokyo

THE EDITOR-IN-CHIEF

John Keith Beddow received his Ph.D. in Metallurgy from Cambridge University, Cambridge, England, in 1959. Currently Secretary of the Fine Particle Society, he is a member of the Faculty at the University of Iowa, where he heads a small research group in fine particle science with emphasis on morphological analysis. Dr. Beddow is an active lecturer and author. He has also been active as a Consultant in metallurgy, powder metallurgy, and powder technology for numerous corporations. He is also president of Shape Technology, Ltd. His present research activities are in particle morphological analysis. Dr. Beddow is married, with four daughters, and has resided in the U.S. since 1966.

THE AUTHOR

Akira Ogawa, Dr. Eng., was born in 1940 in Tokyo and is currently Associate Professor, Department of Mechanical Engineering, College of Engineering, Nihon University, Fukushima, Japan.

He received a Bachelor's degree in Mechanical Engineering from the College of Science and Technology, Nihon University in 1963, a Masters in Engineering from Nihon University in 1965, and a Doctorate in Engineering from Nihon University in 1972.

He worked as an assistant in and as an Assistant Professor at the Department of Mechanical Engineering in the College of Science and Technology, Nihon University from 1968 to 1973. He moved to the Department of Mechanical Engineering, College of Engineering, Nihon University as an Associate Professor in 1974.

Dr. Ogawa is a member of the Japan Society of Mechanical Engineering, the Japan Society of Chemical Engineering, the Japan Society of Powder Technology, the Japan Society of Precision Engineering, the Japan Society of Air Pollution, the Japan Society of Fluid Mechanics, and the Fine Particle Society (U.S.A.).

Dr. Ogawa is the author of *Cyclone Dust Collectors* and *Vortex Flow* (both in Japanese).

His major research interests include the turbulent rotational flow in the confined vortex chamber, the Taylor-Görtler vortex flow in the coaxial cylindrical chamber, turbulent swirling air jets, cyclone dust collectors, rotary flow dust collectors, and the behavior of fine solid particles in turbulent air flow.

ADVISORY BOARD

TABLE OF CONTENTS

Volume I

Volume II

Chapter 5

Chapter 1

CYCLONE DUST COLLECTORS

I. INTRODUCTION

Cyclone dust collectors are one of the most simple dust collectors, do not have movable parts, and are easy to maintain.

A representative illustration of the flow pattern of air, and of the solid particle separation process in the tangential inlet cyclone, is shown in Figure 1.

The centrifugal force

$$Z\left(= \rho_p \cdot \frac{\pi \cdot Xp^3}{6} \cdot \frac{U\theta^2}{r}\right)$$

acting on a solid particle of diameter Xp rotating with velocity Uθ in a circle of a radius r comes to the centrifugal effect (Z/G) = 300 to 2000 in comparison with the gravity force

$$G = \rho_p \cdot \frac{\pi \cdot Xp^3}{6} \cdot g$$

of the solid particle. In spite of the simple construction of cyclones, solid particles have high centrifugal effect value (= Z/G). The mass of the fine solid particle in the centrifugal field apparently is then equivalent to the heavy particle by the centrifugal effect. Therefore, the cyclone dust collectors are useful for separating fine solid particles in the turbulent rotational air flow.

II. TYPES OF CYCLONES

A. Returned Flow Type of Tangential Inlet Cyclone

Figure 2 shows the conical type of a cyclone which was devised by Morse in 1886. This type of cyclone was mainly applied in the separation of the coarse particles or grains.[1]

Figure 3 shows a Linden type of cyclone which was designed by ter Linden in 1935.[2] In order to introduce the dust-laden gas from the periphery into the cyclone by the homogeneous distribution of the dust-laden gas, the inlet pipe is wound over the cyclone body of diameter D1.

ter Linden recommended that the total length L + H be longer than or equal to 3·D1, and D1 is equal to 2.5·D2.

Figure 4 shows the double entry cyclone which divides the inlet dust-laden gas flow into two parts,[3] one of which flows along the outer wall surface and another of which flows in the annular space between the inner pipe and the middle curved plate.

Figure 5 shows a Liot-cyclone which was designed by an institute of the Soviet Union.[4]

B. Axial Flow Cyclone

There are several problems concerning the separation of solid particles in return flow types of the tangential inlet cyclones:

1. When the turbulent rotational flow is returned in the space of the cone, as shown in Figures 1 and 3, the separated dust has the possibility of escaping from the inner pipe with the upward rotational gas flow.
2. The separated dust particles in the dust bunker have a possibility of escaping from the bunker by the secondary flow in the boundary layer due to the remaining turbulent rotational flow in the dust bunker.

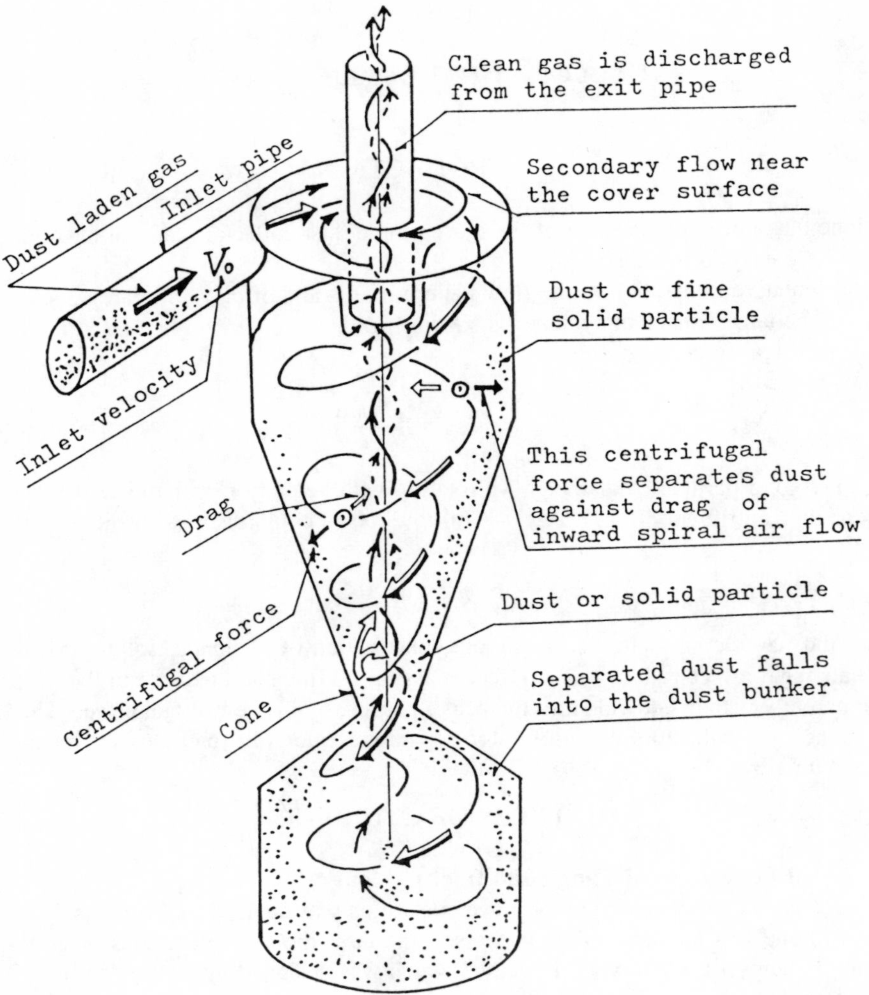

FIGURE 1. Diagram of the flow pattern of the three dimensional spiral flow and of the mechanism of the separation of dust.

In order to correct the above stated defects, the axial flow type of the cyclone was devised. Figure 6 shows the uniflow cyclone which was designed by Umney.[5] The rotational flow is created by the guide vanes. The solid particles are thrown on the outer wall by the centrifugal force in the coaxial space.

The clean gas flows out through the center annular space. This type of cyclone is used in multi-cyclones. The separation efficiency $\eta_c(\%)$ goes down in comparison with the tangential type of the cyclones due to the weak rotational gas flow by means of the guide vanes.

Figure 7 shows a separating unit designed for laminar flow types of cyclones.[6] Radial vanes are set at the center of a cylindrical tube and, assuming that the minimum interval between the inner wall surface of the tube and the outer edges of the radial vanes are d = 12.5 mm or one tenth the diameter of the separating tube of interval diameter of D1 = 125 mm, and also assuming that the gas moves through a separating unit with Vo = 1.5 m/s at a temperature T = 473 K, then the flow Reynolds number Re may be estimated as

$$Re = \frac{Vo \cdot d}{\upsilon} \doteqdot 5400$$

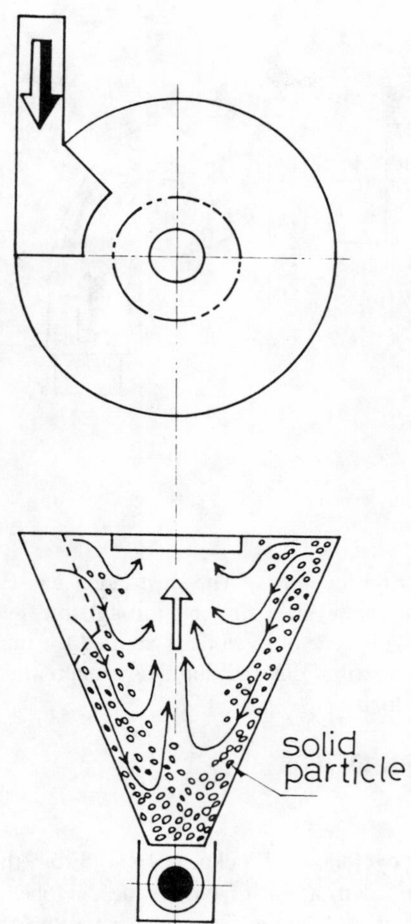

FIGURE 2. Morse-designed cyclone (1886).

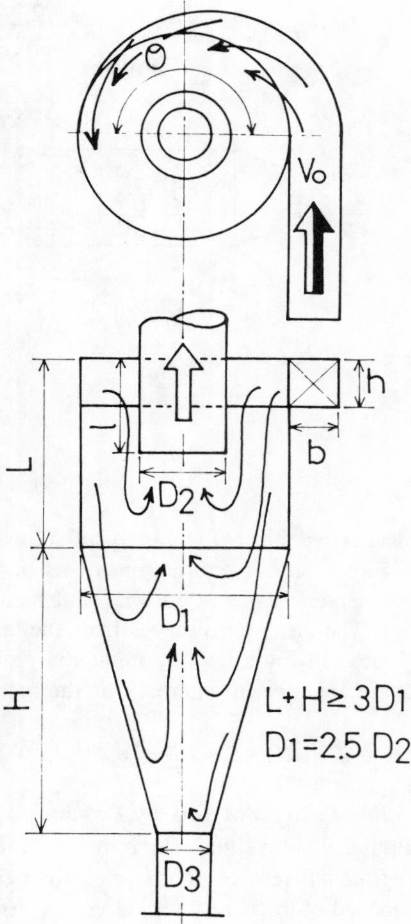

FIGURE 3. Linden cyclone (1935).

Therefore, under these flow conditions, the gas flow may be assumed to be stabilized in a laminar condition as long as the Reynolds number is below approximately 10,000 in centrifugal type separating units[7] (British Patent 713,930). However, according to the hydrodynamical consideration for quasi-free vortex flow in this interval, this type of vortex flow has a tendency to become an unstable flow. Therefore, even if the straight flow may be assumed to be a laminar flow, the spiral flow which is combined with straight flow and vortex flow becomes an unstable flow condition.

A particular advantage of this radial vane form is that the spiral flows are set up by the frictional drag of the main stream rotating in the annular zone around the core in each of the four quadrants of the cross-shaped vanes. The eddy in each quadrant rotates at its outer surface in the same direction as the main flow so that there is little, if any, frictional drag between the main stream and the localized eddy in each quadrant. The effect is to cause the main gas stream to roll around and on the eddy as if rolling on roller bearings.

The result is stability of the main gas flow with a minimum of friction or energy dissipation by virtue of contact with the radial vanes. This fact lends stability to the main gas flow and helps maintain its laminar character. Any dust or solid particles that enter into these rotational flows are thrown out in a short time because the centrifugal force applied to them while in the rotational flow is relatively high.

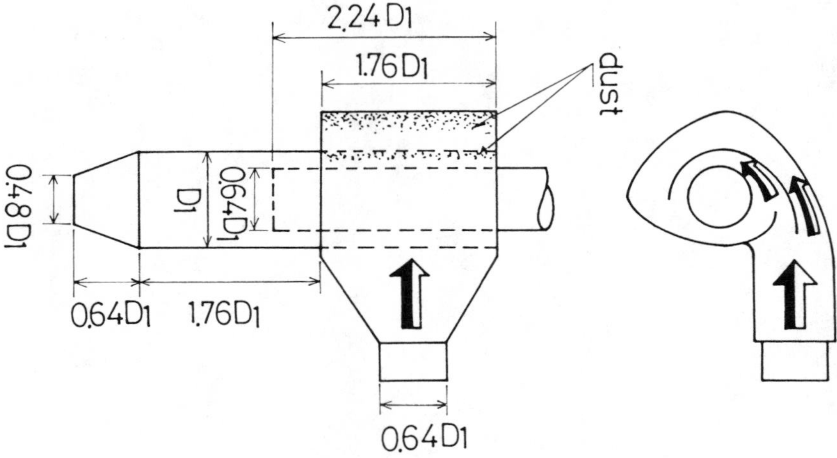

FIGURE 4. Double entry cyclone.

C. Returned Flow Type of Axial Flow Cyclone

Figure 8 shows return flow types of the axial flow cyclones. The dust-laden gas flow axially enters into the cyclone and then begins to rotate after flowing through the guide vanes. The clean gas escapes from the inner pipe. This type of cyclone is applied for multi-cyclones. In Figure 8, the tangential velocity Vo distributions measured by Jotaki[8] in the cyclone are shown, where Vo is the mean inlet velocity.

III. FLOW PATTERN OF GAS

A. Flow Pattern of Gas by Prockat

Figure 9 shows the velocity distributions in the cyclone by Prockat (1930).[9] From these experimental results, it was clear that the velocity distribution of gas in the cyclone was composed of the nearly forced vortex flow in the central region and of the nearly free vortex flow in the outer region.

B. Secondary Flow Pattern

Figure 10 shows the observation of the secondary flow pattern by Hughes (1957).[10] It is very important to recognize that the flow of the short circuit near the surface of the inner pipe occurs by the boundary layer effect. This short circuit remarkably drops the collection efficiency of the dusts.

Figure 11 shows the very interesting double vortex flow in the van Tongeren type of cyclone. This flow pattern was observed by Jackson (1962).[11]

Figure 12 shows the secondary flow pattern in the annular space near the inlet pipe region. Those stream lines were calculated by Liu, Jia, Zhang, Hao, Wang, and Xu (1978).[12]

Figure 13 shows the secondary flow patterns near the inlet pipe region around the space between the outer pipe D1 = 150 mm and the inner pipe D2 = 51 mm in a cylindrical cyclone.[13] The mean inlet velocity Vo in the inlet pipe is Vo = 20 m/s. From this figure, it can be seen that the flow pattern of the secondary flow in the cyclone dust collector is not always the same flow pattern, but changes the flow pattern from the inlet angle θ = 0.89 rad to the transverse angle θ = 3.14 rad. This flow pattern is very important in considering the collection efficiency of fine solid particles.

C. Tangential Velocity Distribution

1. Rankine Combined Vortex

One of the simplest vortex models is Rankine's combined vortex flow which is composed of the forced vortex and the free vortex. The forced vortex can be represented by

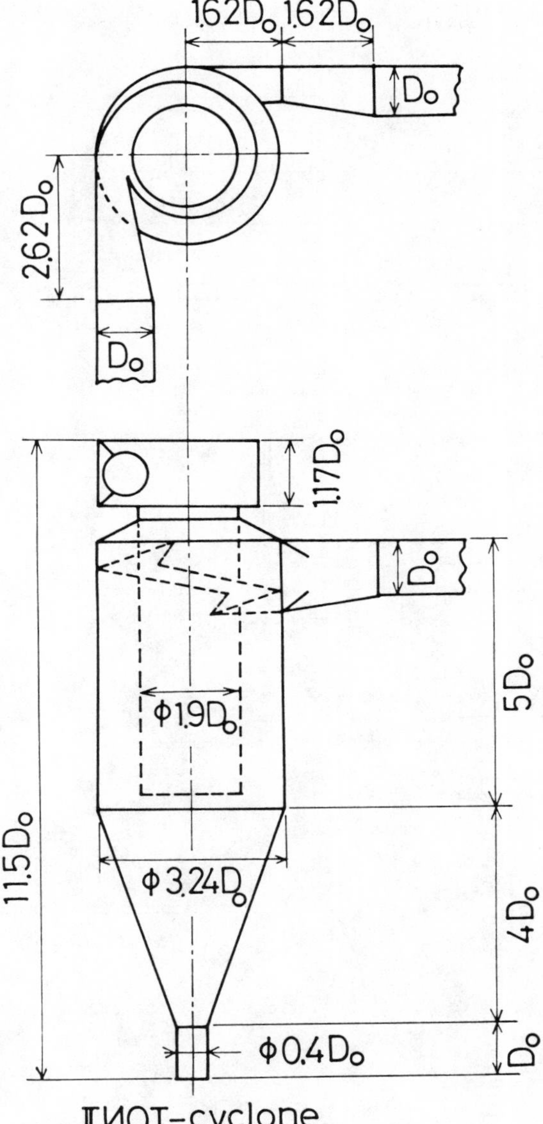

FIGURE 5. Liot-cyclone.

$$V\theta = \omega \cdot r \; (0 \leqq r \leqq a) \tag{1}$$

and the free vortex can be written

$$V\theta \cdot r = \Gamma \; (r \geqq a) \tag{2}$$

where a symbol ω(rad/s) is an angular velocity and a symbol Γ(m^2/s) means a circulation. This vortex model is shown in Figure 14. The static pressure distributions can be represented as:

$$p = p_a + \frac{\rho \cdot \Gamma^2}{2} \cdot \left(\frac{1}{a^2} - \frac{1}{r^2} \right) \tag{3}$$

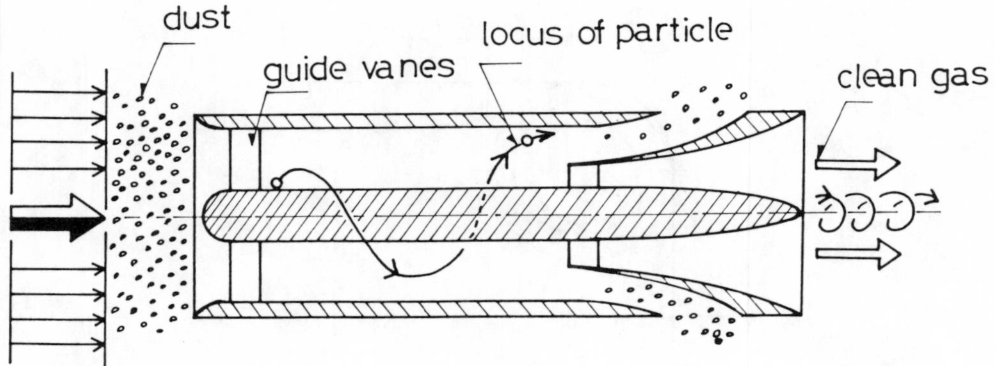

FIGURE 6. Umney cyclone (1948).

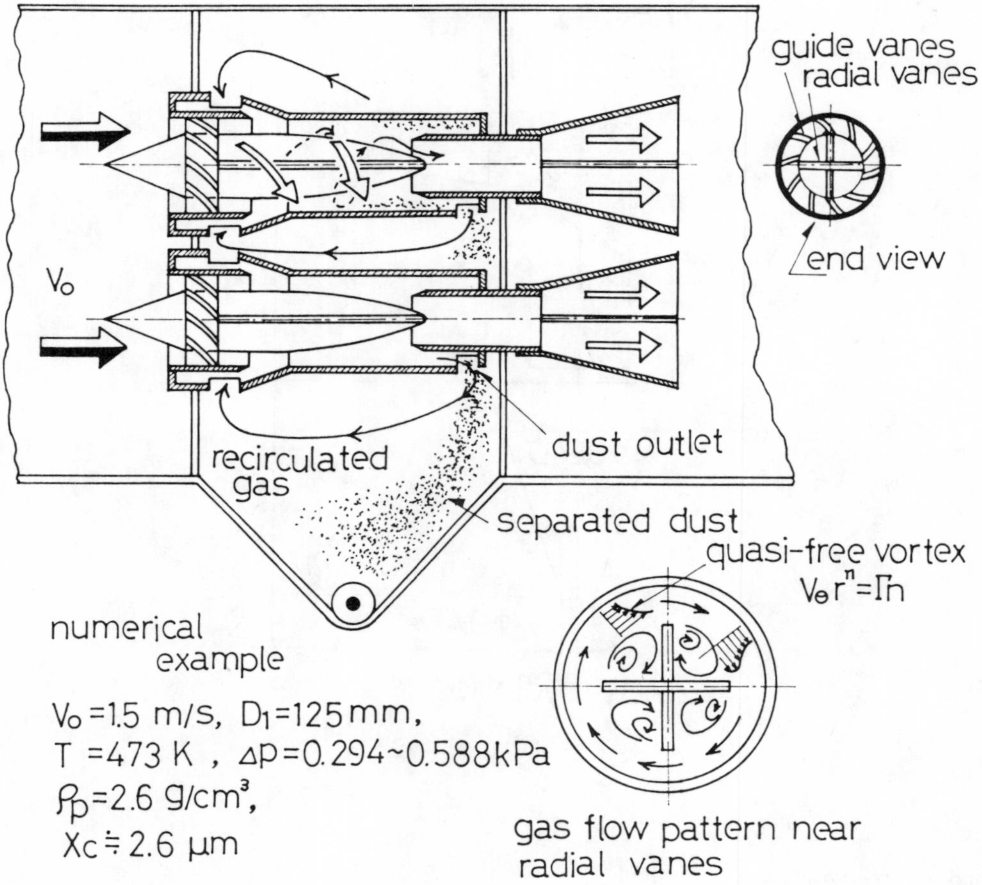

FIGURE 7. Laminar flow cyclone.

and for the free vortex

$$p = p_0 + \frac{\rho \cdot \Gamma^2}{2 \cdot a^4} \cdot r^2 \tag{4}$$

Therefore, the difference of the static pressures $(0 - p_a)$ and $(P_a - P_o)$ is equal to

$$0 - p_a = p_a - p_0 = \frac{\rho \cdot \Gamma^2}{2 \cdot a^2} \tag{5}$$

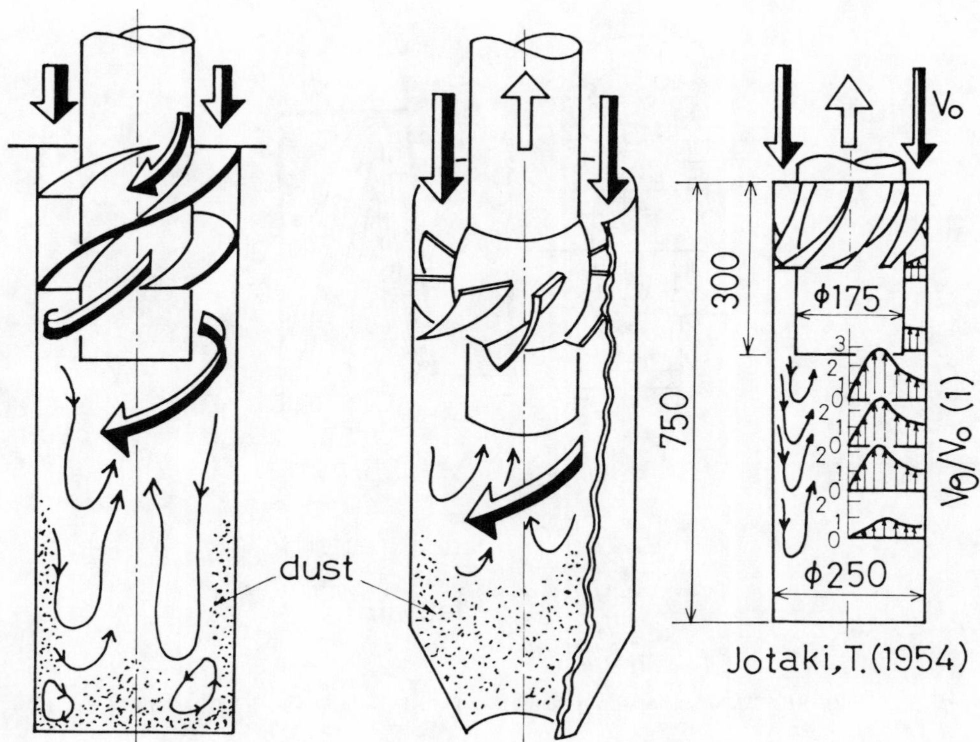

FIGURE 8. Return flow types of the axial flow cyclones.

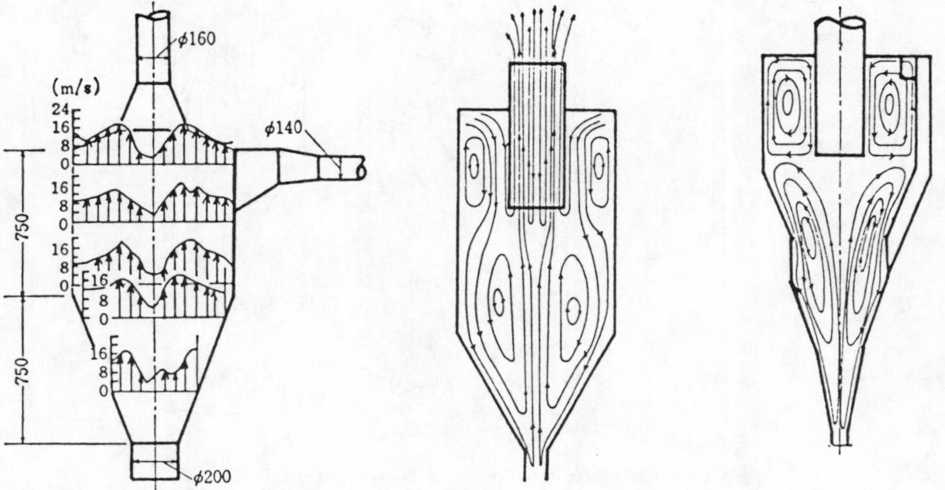

FIGURE 9 (left). Velocity distributions in the cyclone by Prockat (1930). FIGURE 10 (center). Secondary flow in the cyclone by Hughes (1957). FIGURE 11 (right). Double vortex flow in the cyclone by Jackson (1962).

where 0 means atmospheric pressure.

2. *Burgers' Vortex Model*

In the turbulent rotational flow, namely, a Reynolds stress larger than a viscous stress, Burgers (1948)[14] derived his vortex model by using the Navier-Stokes equation and the

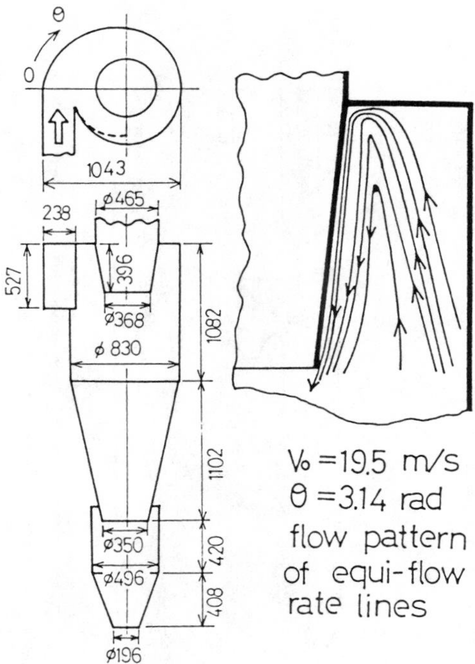

FIGURE 12. Secondary flow pattern (Data from *Acta Mech. Sin.*, 3, 182, 1978.).

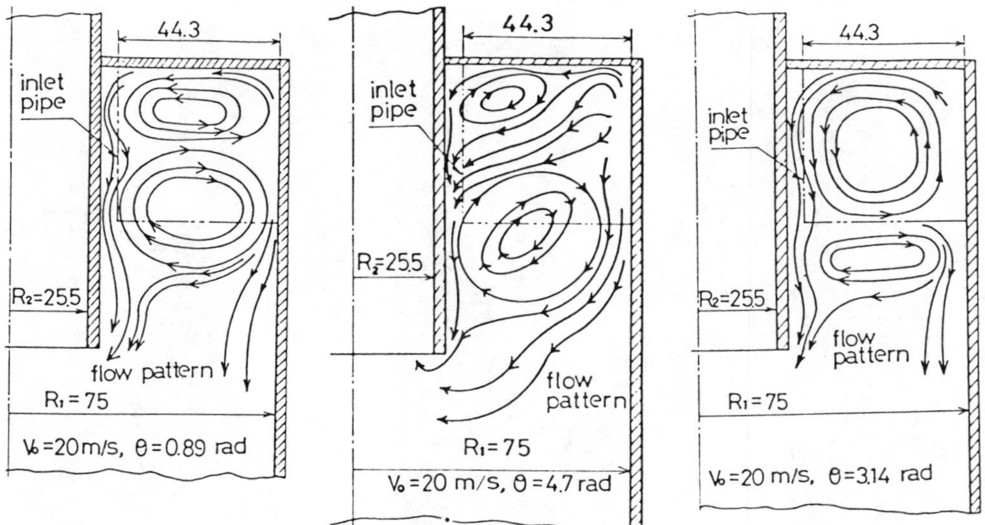

FIGURE 13. Secondary flows near the inlet pipe region.

continuity equation. Denoting that the tangential and radial velocities at the radius r = b are Vθb and Vrb, respectively, then the dimensionless radius r′ = r/b, the dimensionless velocities Vθ′ = Vθ/Vθb, and then Vr′ = Vr/Vrb are introduced, so the equations of the tangential and radial velocities in the radius region $0 \leqq r \leqq b$ can be represented as

$$V'_\theta = \frac{1}{r'} \frac{1 - \exp(-\,Reb \cdot r'^2/2)}{1 - \exp(-\,Reb\,/2)} \qquad (6)$$

$$V'_r = r' \qquad (7)$$

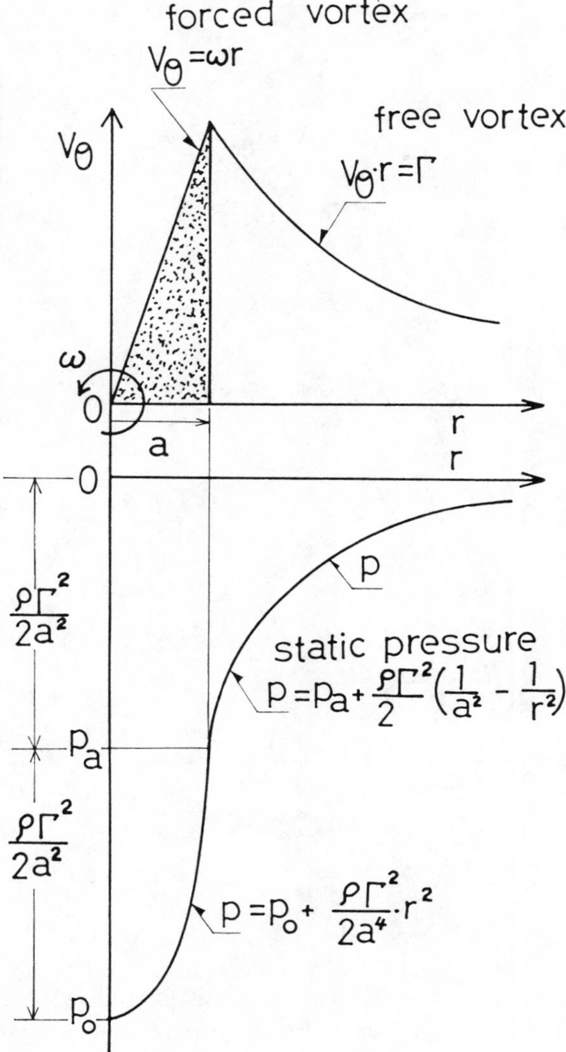

FIGURE 14. Rankine combined vortex model.

where the dimensionless symbol Reb means a Reynolds number defined as

$$\text{Reb} = \frac{-\,b \cdot Vrb}{\upsilon_T} \tag{8}$$

The symbol υ_T(m^2/s) means an eddy kinematic viscosity.

Distribution curves of Equation 6 of Burgers' vortex model are shown in Figure 15 for the various kinds of the Reynolds number Reb. In the figures, the symbols ○,△,□,●,▲ are the measured tangential velocities in the cyclone of D1 = 150 mm, D2 = 50 mm, and Do = 50 mm. Figure 16 shows the velocity distribution of Burgers' vortex model for Reb = 20. The measured velocities are shown for D1 = 150 mm, D2 = 50 mm, and Do = 50 mm. In addition, the velocity distribution near the outer wall surface measured by Muschelknautz and Krambrock (1970) is shown.

Figure 17 shows the distributions of the eddy kinematic viscosity for the inlet velocity of

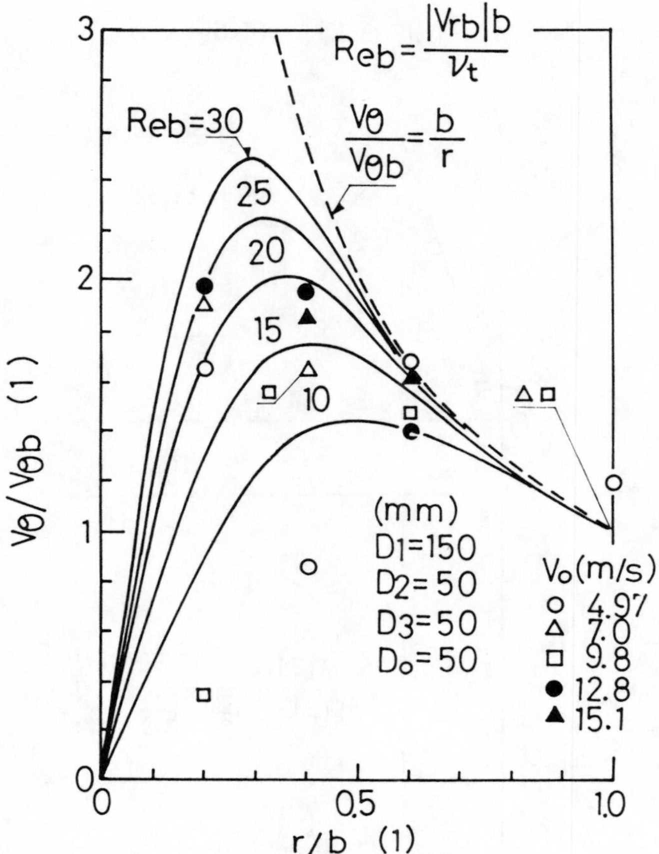

FIGURE 15. Burgers' vortex flow distributions.

air Vo = 5, 7, 10, 13, and 15 m/s in the vortex chamber of D1 = 141 mm and $H_T \doteqdot 380$ mm. Those values of υ_T/υ were measured with the hot-wire of X-probe by Ogawa and Fujita (1981). The maximum value of υ_T/υ is located in the quasi-forced vortex region.

Figure 18 shows Prandtl mixing length l in the vortex chamber of Figure 17. In order to estimate the mixing length l, the following equation was applied by Ogawa and Fujita defined as

$$\tau_T = -\rho \cdot \overline{v_r \cdot v_\theta} = \rho \cdot l^2 \cdot \left(\frac{dV\theta}{dr} - \frac{V\theta}{r}\right)^2 \tag{9}$$

From the results of the mixing length of Figure 18, Ogawa (1981) derived the equation of the mixing length l in the confined vortex chamber as

$$\frac{l}{R1} = 6 \times 10^{-6} \cdot Rcy \cdot \frac{r}{R1} \tag{10}$$

where a symbol Rcy is the cyclone Reynolds number defined as

$$Rcy = \frac{Qo}{Hi \cdot \upsilon} \tag{11}$$

Qo(m^3/s) is the flow rate of gas into the vortex chamber and Hi(m) is the effective length of the imaginary cylinder in the vortex chamber or cyclone.

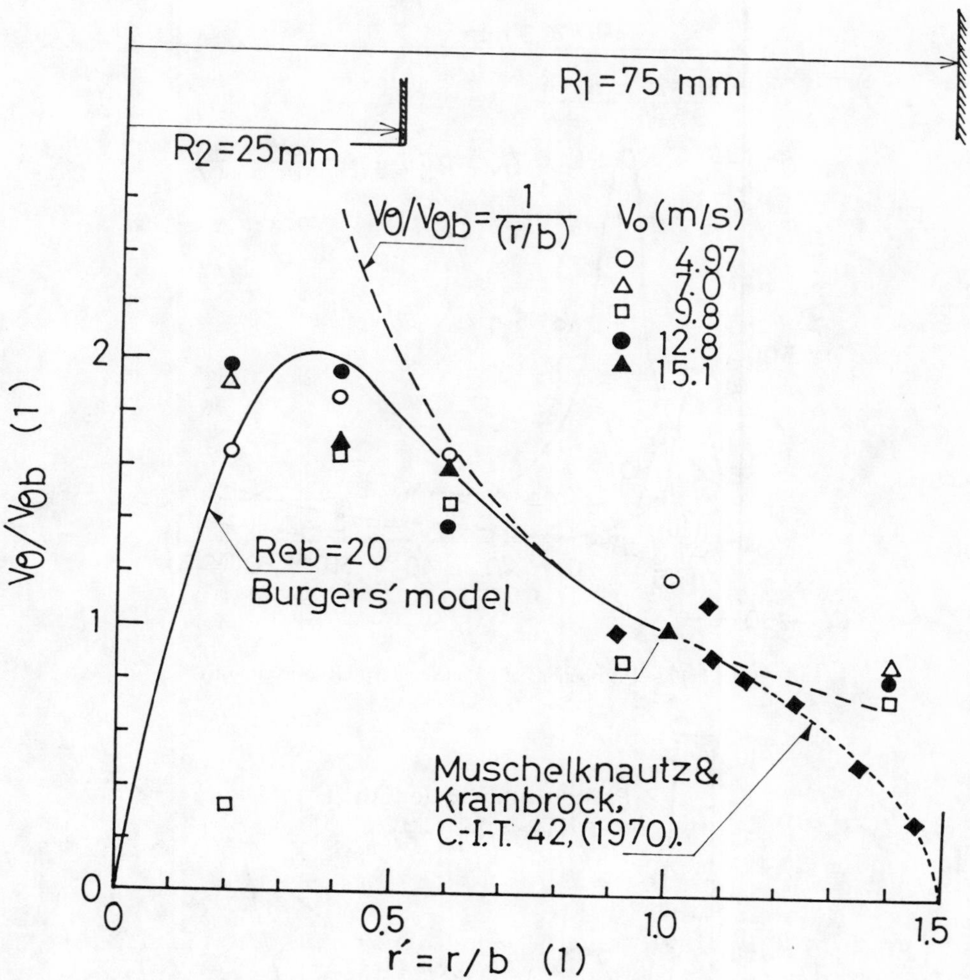

FIGURE 16. Tangential velocity distributions by Burgers vortex flow and Muschelknautz and Krambrock.

3. Ogawa's Combined Vortex

Ogawa's combined vortex model is composed of the quasi-forced vortex ($0 \leqq r \leqq r_t$) and quasi-free vortex ($r \geqq r_t$) defined as

$$V\theta = K \cdot r \cdot \exp(-\Lambda r) \doteqdot K \cdot r(1 - \Lambda \cdot r), \quad Vr = -f \cdot r \tag{12}$$

for the quasi-forced vortex

$$V\theta \cdot r^n = \Gamma n, \quad Vr = -\frac{m}{r} \tag{13}$$

where the symbols f(1/s) and m(m²/s) are constant values and the relationship between Λ, r_t, a, and n are written

$$\Lambda \cdot r_t = \frac{1 + n}{2 + n}, \quad a \cdot \Lambda = \frac{1}{2} \tag{14}$$

Then a symbol r_t (m) means the boundary radius between the quasi-forced and quasi-free vortices, and n(1) means the velocity index. The value of n is equal to 1 for the potential

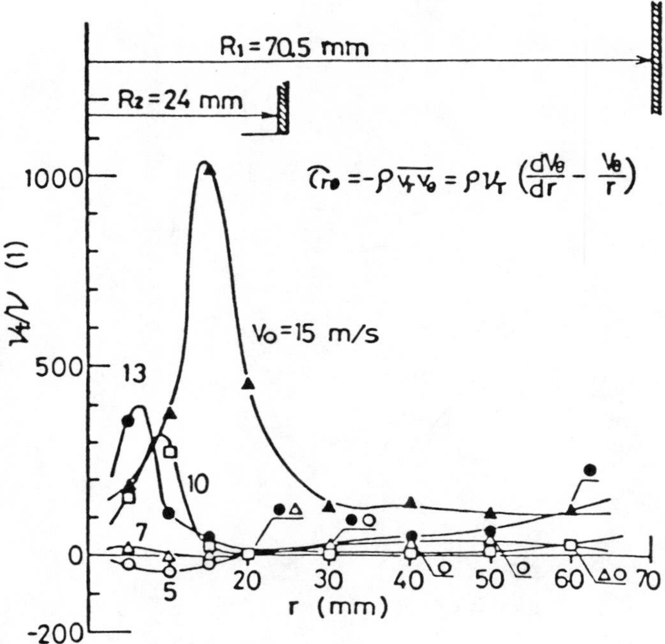

FIGURE 17. Distributions of turbulent eddy kinematic viscosity.

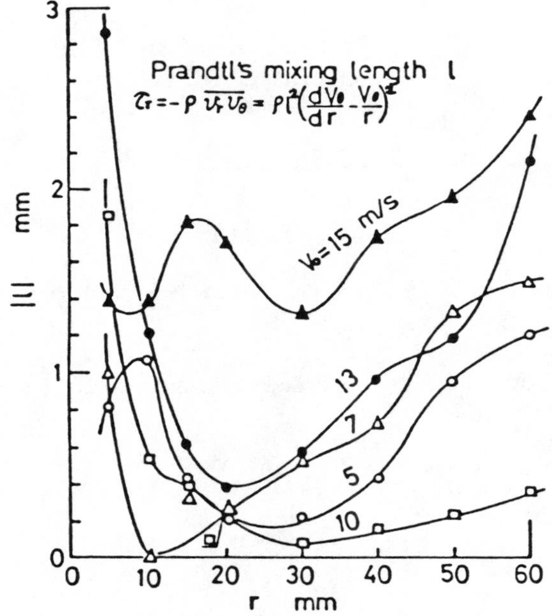

FIGURE 18. Distributions of mixing length.

flow and the value of n decreases with increasing the effect of the gas viscosity. Figure 19 shows an illustration of Ogawa's combined vortex model.[15]

Figure 20 shows the experimental value of the velocity index n for the various types of cyclones and for the vortex chamber. In this figure, Qb(m^3/s) means the flow rate into the dust bunker of the cyclone. The equation of the flow rate Qb was derived by Ogawa as follows:

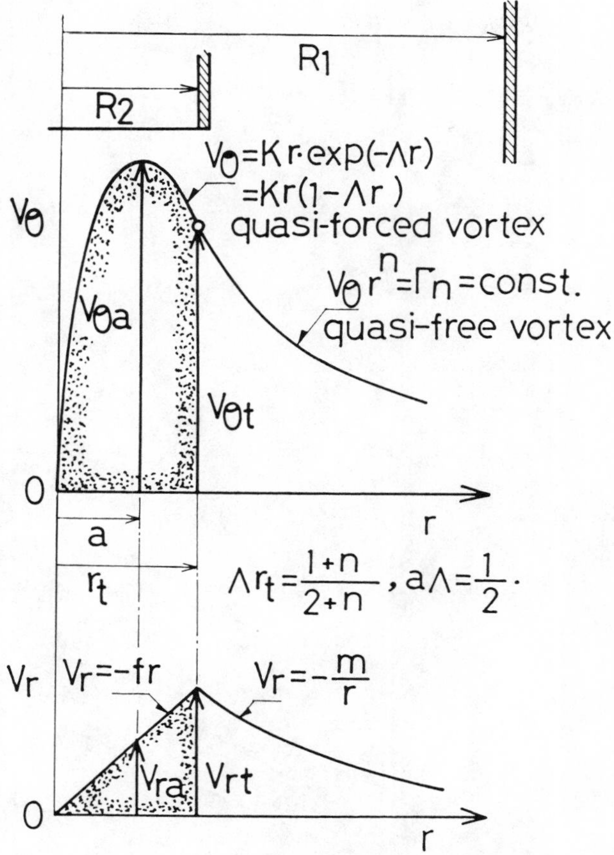

FIGURE 19. Ogawa's combined vortex model.

$$\frac{Qb}{Qo} = 2.62\left\{1 - 0.36\left(\frac{D3}{D1}\right)^{-0.56}\right\}^{3/2} \cdot \left(\frac{D3}{D1}\right)^{1.16} \qquad (15)$$

From this figure, you will find that the value of n increases with increasing the cyclone Reynolds number Rcy.

On the other hand, concerning the value of K(1/s) of Ogawa's combined vortex model, Hikichi (1982) found the characteristics of K, which was a function of diameter D2 of cyclone and flow rate Qo and also independent of diameter D1 as shown in Figure 21. Table 1 describes the main sizes of the cyclones. Ogawa obtained the empirical equation of K

$$K(1/s) = 66.5 \cdot D2^{-2.65} \cdot Qo \qquad (16)$$

where D2(m) is a diameter of the inner pipe and Qo(m^3/s) is the flow rate into the cyclone. This equation is very important for the estimation of Ogawa's combined vortex model.

IV. PRESSURE DROP

A. Definition of the Pressure Drop

1. Suction Type

As shown in Figure 22, generally speaking, the pressure drop Δp_c by the turbulent rotation in the cyclone can be described by the pressure difference between the mean total pressure

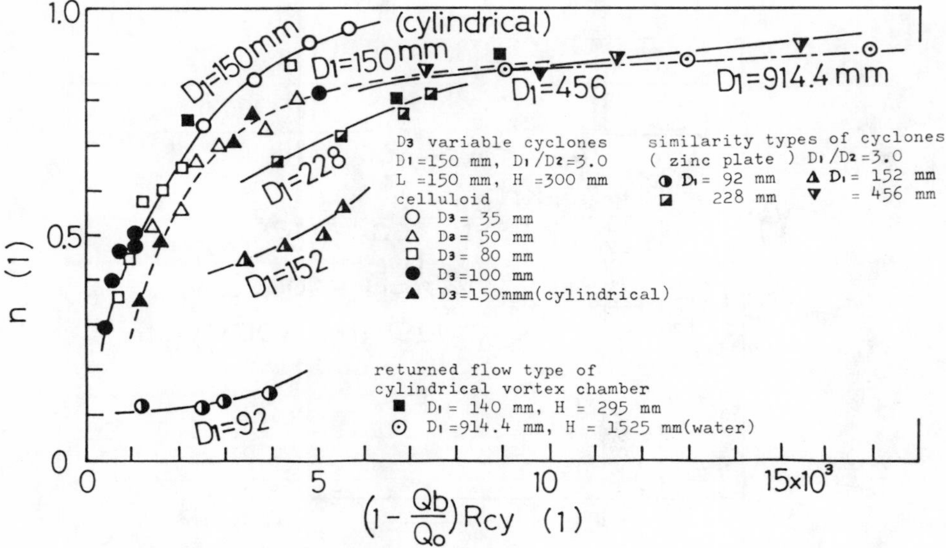

FIGURE 20. Velocity index for the various types of the cyclones and of the vortex chambers by Hikichi.

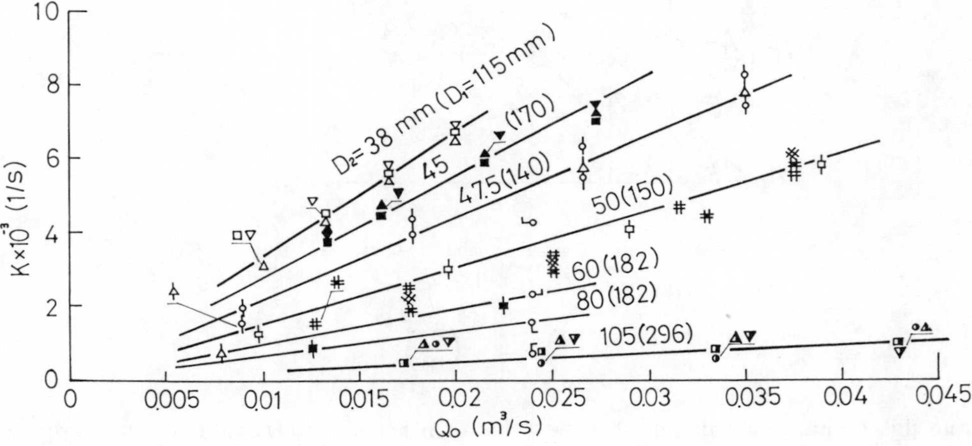

FIGURE 21. Characteristics of K of Ogawa's combined vortex model by Hikichi.

P_T1 (static plus dynamic pressures) in the inlet pipe and the mean total pressures P_T2 in the inner pipe as follows:

$$\Delta p_c = P_T1 - P_T2 = P_{s1} + \rho \cdot \frac{Vo^2}{2} - p_{s2} - \rho \cdot \frac{\overline{(V\theta^2 2 + Vz^2 2)}}{2} \qquad (17)$$

where Vθ2 and Vz2 are the tangential and axial velocities of gas in the inner pipe and also $\overline{V_\theta^2 2 + V_z^2 2}$ means the mean dynamic pressure in the inner pipe.

Here we must turn our attention to the measurement of the static pressure p_{s2} at the exit pipe (inner pipe). The static pressure distribution is a higher pressure at the inner pipe wall and a lower pressure at the pipe's center due to the kinetic energy of the fluid in the inner pipe. Therefore, when we measure the difference $(p_{s1} - p_{s2})$ of the static pressures by the

Table 1
MAIN SIZES OF CYCLONE DUST COLLECTORS

Symbol	Size
* ZB (D_1 =115 mm), D_2/D_1 = 0.33	
△	Z= 75mm
□	Z=135mm
▽	Z=192mm
* CIII (D_1 =296 mm), D_2/D_1 =0.355	
◑	Z= 50mm
◭	Z=100mm
◧	Z=250mm
▼	Z=492mm
* CI (D_1 =170mm), D_2/D_1 =0.265	
▲	Z= 54.5 mm
▼	Z=104.5mm
■	Z=204.5mm
** (D_1 =150mm), D_2/D_1 =0.33	
⌑	Z= 30mm
*** (D_1 =182mm), D_2/D_1 =0.56	
δ	Z= 29mm
*** (D_1 =182mm), D_2/D_1 =0.43	
♀	Z= 69mm
*** (D_1 =182mm), D_2/D_1 =0.33	
o⅃	Z=289mm
*** (D_1 =182mm), D_2/D_1 =0.28	
└o	Z=219mm

Symbol	Size
** (D_1 =150 mm), D_2/D_1 =0.33	
※	Z= 50mm, D_3 = 35mm,
#	Z= 50mm, D_3 = 50mm,
⋕	Z= 50mm, D_3 =100mm,
** (D_1 =140 mm), D_2/D_1 =0.34	
⚲	Z= 43mm,
⍋	Z=262mm,
** (D_1 =290 mm), D_2/D_1 =0.17	
⧯	Z=405mm,

* Hikichi, T., Ogawa, A.,

** Ogawa, A.,

*** Iinoya, K.,

manometers, we read the lower pressure (about 50% of the true pressure differences) in comparison with the true pressure differences. Then we call the measured pressure differences

$$\Delta p_{ca} = p_{s1} - p_{s2}$$

as an apparent pressure drop.

Consequently, if we assume that the area of the inlet pipe Ao(m^2) is equal to the area of the inner pipe A2(m^2), so the cyclone pressure drop may be written nearly as

$$\Delta p_{ca} \doteqdot p_{s1} - p_{s2} \tag{18}$$

For this reason, when we choose the type of blower based upon the apparent pressure drop Δp_{ca}, so the collection efficiency η_c of the cyclone falls remarkably due to the weak rotational velocity of the fluid.

2. Pressure Type

One of the most simple methods for measuring the true pressure drop Δp_c of a cyclone

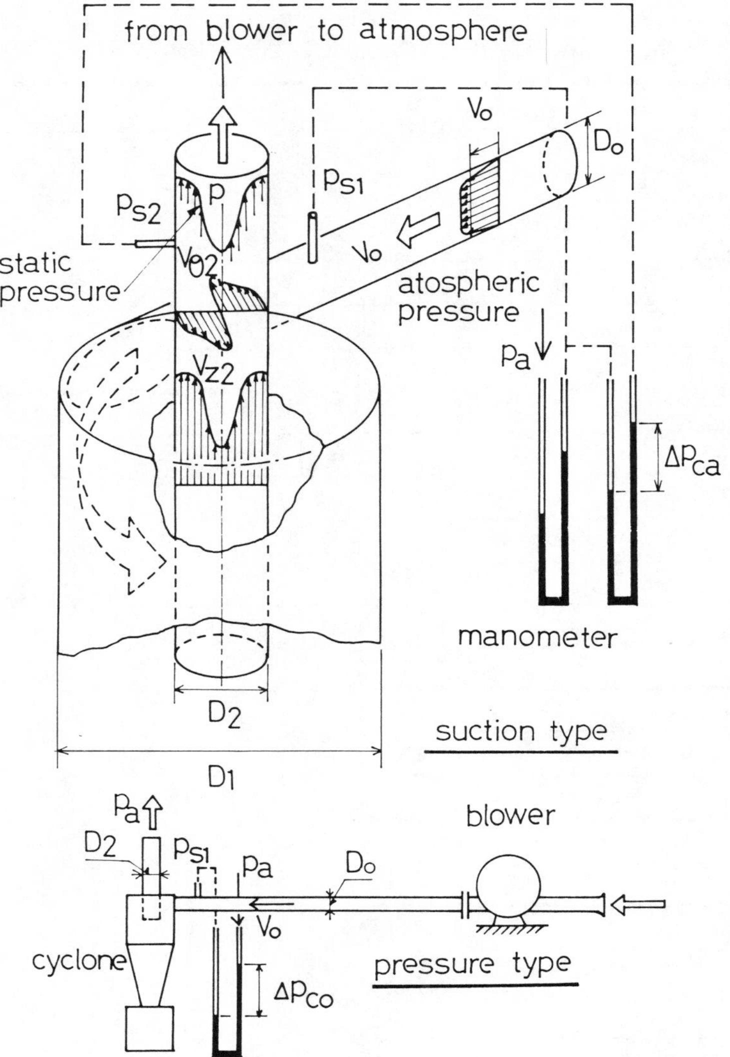

FIGURE 22. Illustrations of the measurement method of the pressure drop.

by using the pressure type system, is shown in Figure 22. In this method, the exit area of the inner pipe is open so the rotational air flow leaving the inner pipe is diffused into the atmosphere. Therefore, the mean kinetic energy $\rho\cdot(\overline{V_\theta^2 2 + V_z^2 2})/2$ of the rotational fluid becomes zero in the atmosphere.

Further, the static pressure p_{s2} at the inner pipe is equal to the atmospheric pressure p_a, so the true pressure drop Δp_c can be written as follows:

$$\Delta p_c = (p_{s1} - p_a) + \rho\cdot\frac{V_o^2}{2} = \Delta p_{co} + \rho\cdot\frac{V_o^2}{2} \tag{19}$$

B. Empirical Equation of the Pressure Drop

Generally speaking, about 80% of the cyclone pressure drop is based upon the energy dissipation by the viscous stress of the turbulent rotational flow. About 20% of that is based upon the pressure drop by the fluid friction on the surface of the cyclone and by the sudden expansion and contraction of fluid flow in the cyclone.

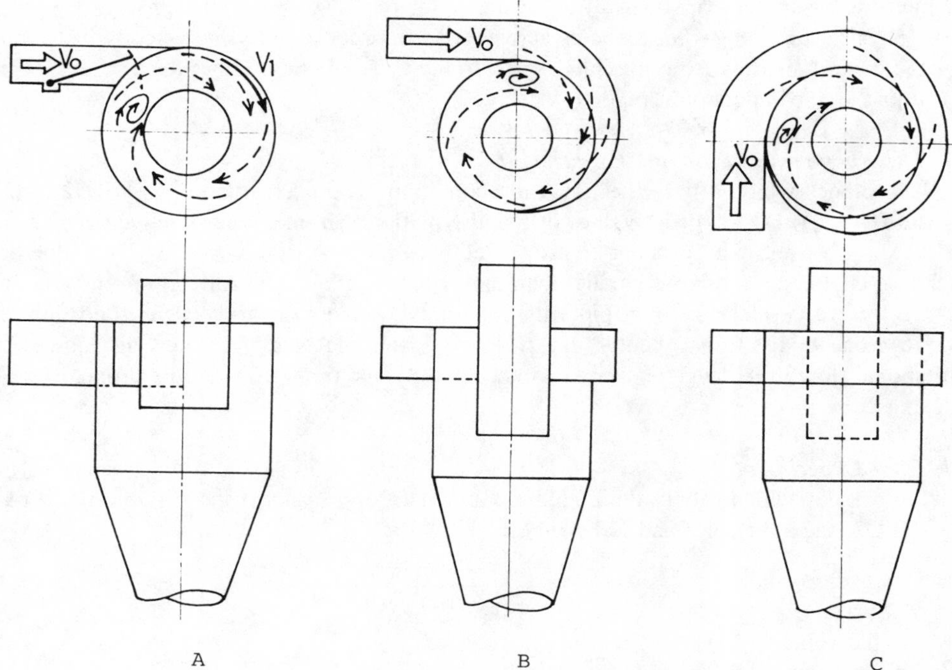

FIGURE 23. Fundamental types of cyclone dust collectors.

The empirical equation of the pressure drop Δp_c can be defined as

$$\Delta p_c = \zeta_c \cdot \rho \cdot \frac{Vo^2}{2} \tag{20}$$

where Vo(m/s) is the mean inlet velocity in the inlet pipe and ζc(1) is called the coefficient of the pressure drop in the cyclone and refers to the mean dynamic pressure ($\rho \cdot Vo^2$) in the inlet pipe.

Further, the coefficient of the pressure drop is defined as

$$\zeta_c = X \cdot \left(\frac{Ao}{D2^2}\right)^z \tag{21}$$

where X and z are functions of the cyclone geometries and describe the various kinds of the cyclones.

1. Linden Experiment

In the ordinary tangential inlet type of cyclone, as shown in Figure 23, without setting up the control inlet valve of type (A), the values of z and X become z = 1.0, X = 25 to 32, and in the type (B) of the 180° spiral inlet pipe as shown in Figure 23, the value of z and X become z = 1.0, X = 21.7. Here the value of X is for the 180° spiral inlet pipe (Figure 23). This phenomenon is caused by the insufficient total energy of the feed gas in the cyclone due to the additive energy loss of surface friction of the fluid while flowing through the spiral inlet pipe.

2. Lapple Experiment

In the 360° spiral inlet pipe type which is enveloped on a cyclone body, the values of z and X become z = 1.0, X = 18.4. Here, the value of X for the 360° spiral inlet pipe shows a lower value in comparison with the value X for the 180° spiral inlet pipe of Figure 23. The physical meaning is the same as above. Also in order to unify the velocity distribution of the inward air flow from the inlet pipe into the cyclone body, either the 180° inlet spiral pipe or the 360° inlet spiral pipe are designed.

3. The Shepherd and Lapple Experiment

In cyclones which satisfy the following condition of the construction as D1/D2 = 2 to 4 and (L + H)/D2 = 4, the values of z and X without an inlet control valve become z = 1.0, X = 16, and with an inlet control value become z = 1.0, X = 7.5. The value of X with an inlet control valve is smaller than that without an inlet control valve. When the inlet control valve is set in the inlet pipe, the inlet valve hinders remarkably the fluid flow into the cyclone body. Consequently, the fluid dissipates the energy by making eddies. The rotational flow in the cyclone decreases and, at the same time, the pressure drop Δp_c is also decreased.

4. First Experiment

In cyclones which satisfy the conditions of D1/D2 = 2 and of the variable value of (L + H)/D2, the value of z and X becomes

$$z = 1.0, \quad X = \frac{(12/Y)}{(L/D1)^{1/3} \cdot (H/D1)^{1/2}}$$

where the value of Y is

without inlet valve, Y = 0.5
middle position of the inlet valve in the inlet pipe, Y = 1.0
complete closed position of the inlet valve, Y = 2.0

5. Alexander Experiment

Alexander derived the equation of the pressure drop based upon the pressure drop in the quasi-free vortex region in the cyclone and the pressure drop of the rotational flow flowing through the inner pipe. Then he determined the values of z and X as

$$z = 1.0$$

$$z = 4.62\left(\frac{D2}{D1}\right)\left[\left\{\left(\frac{D1}{D2}\right)^{2n} - 1\right\}\cdot\left(\frac{1-n}{n}\right) + f\left(\frac{D1}{D2}\right)^{2n}\right]$$

where the value of f is changed with the value of the velocity index n. The relationship between f and n is shown in Figure 24. The relationship between X and D1/D2 is shown with a parameter n in Figure 25.

6. Stairmand Experiment

Stairmand derived the following equation of the coefficient of pressure drop as

$$\zeta_c = K3 + 3.24\cdot\left(\frac{Ao}{D2^2}\right)^2$$

$$K3 = 1 + 2\left(\frac{V1}{Vo}\right)^2 + \left\{\left(\frac{D1 - Do}{D2}\right) - 1\right\}$$

where a symbol V1 means the tangential velocity of fluid along the surface of the outer pipe in the cyclone body. Then the relationship between K3 and (D1/D2) is shown in Figure 26. In this figure, λ means a friction factor and S means the total area of the cyclone surface. The relationship between V1/Vo and $\sqrt{Ao}/D1$ may be represented by the following values:

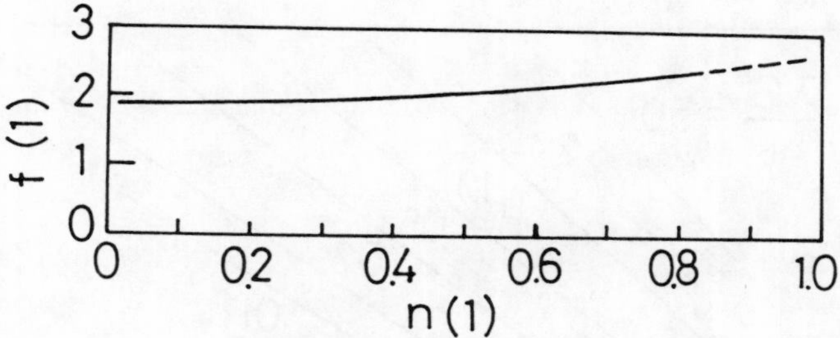

FIGURE 24. Relationship between f and n.

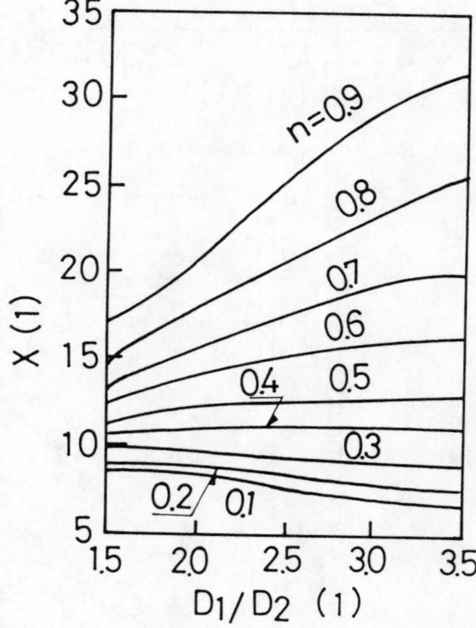

FIGURE 25. Relationship between X and D1/D2.

$$\sqrt{Ao/D1} > 0.35 \qquad V1/Vo = 1.4$$

$$0.35 > \sqrt{Ao/D1} > 0.20 \qquad V1/Vo = 4 \cdot \sqrt{Ao/D1}$$

$$\sqrt{Ao/D1} < 0.20 \qquad V1/Vo = 0.8$$

7. Casal and Martinez Experiment

Casal and Martinez (1981)[16] derived an empirical equation of the pressure drop coefficient by means of a statistical analysis method.

The pressure drop Δp_c (Pa) of the cyclone is defined as

$$\Delta p_c = \zeta_c \cdot \rho \cdot \frac{Vo^2}{2} \tag{22}$$

where $\zeta c(1)$ is the coefficient of the pressure drop. They determined ζc as

$$\zeta_c = 11.3\left(\frac{a \cdot b}{D2}\right)^2 + 3.33 \tag{23}$$

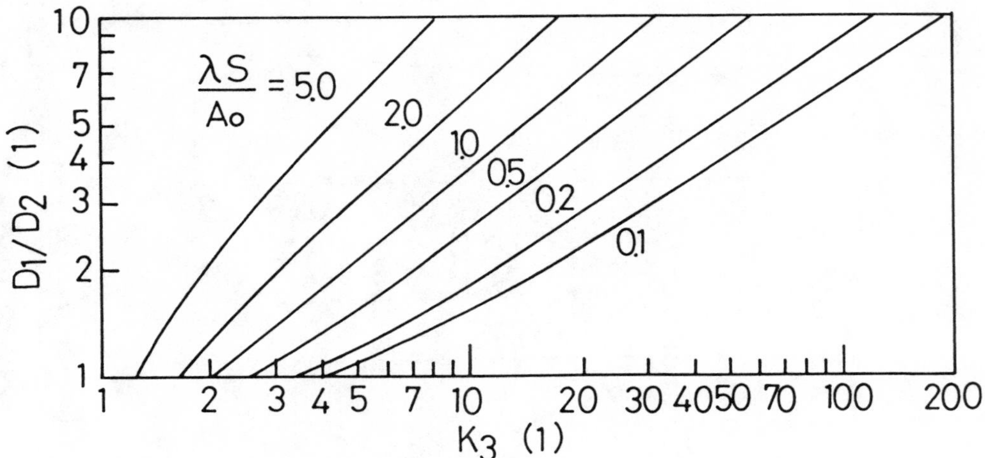

FIGURE 26. Relationship between D1/D2 and K3.

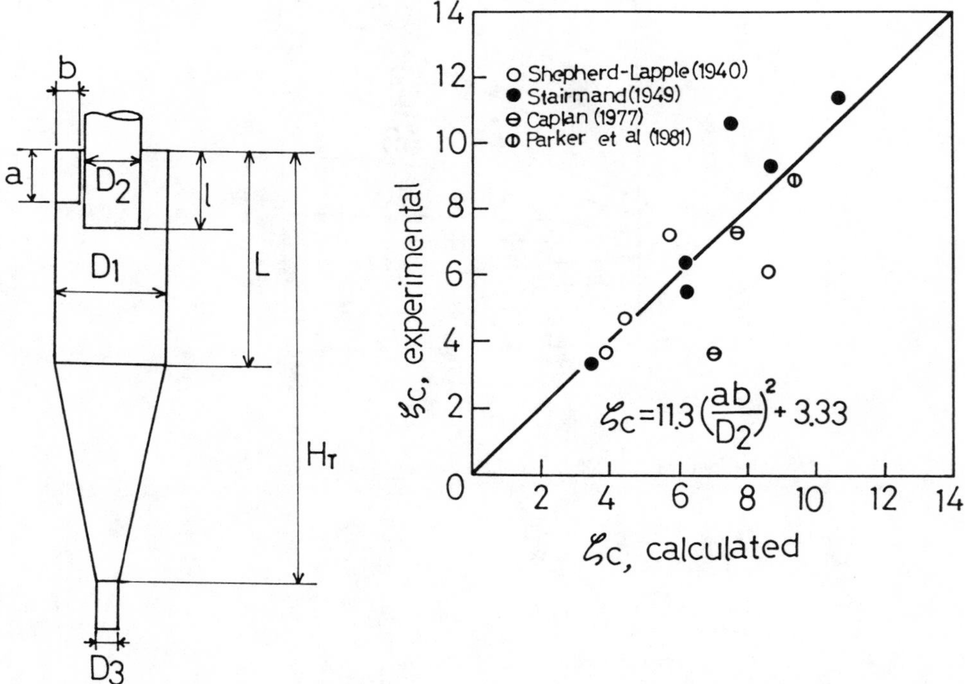

FIGURE 27. Empirical equation of the coefficient of the pressure drop.

The relationship between the experimental values of ζc and the calculated values ζc of Equation 23 is shown in Figure 27.

8. Iinoya Experiment

Iinoya derived the empirical equation of ζc as

$$\zeta_c = K \cdot \left(\frac{Ao}{D2^2}\right) \cdot \sqrt{\frac{D1}{L + H}} \tag{24}$$

where a dimensionless factor K is, in general, 20 to 40 and depends on the shape of the inlet pipe as

1. In a circular pipe of diameter Do

$$\zeta_c = 0.785 \cdot K \cdot \left(\frac{Do}{D2}\right)^2 \cdot \sqrt{\frac{D1}{L + H}} \tag{25}$$

2. In a rectangular inlet pipe of width b and height h

$$\zeta_c = K \cdot \left(\frac{b \cdot h}{D2^2}\right) \cdot \sqrt{\frac{D1}{L + H}} \tag{26}$$

9. Harada and Ichige Experiment

Harada and Ichige derived an empirical equation for small sized cyclones.

1. In the circular inlet pipe of diameter Do,

$$\zeta_c = 6.05 \cdot \left(\frac{Vo \cdot Do}{\upsilon}\right)^{0.12} \cdot \frac{Ao^{0.94} \cdot D1^{0.12}}{D2^{1.6} \cdot Hi^{0.25} \cdot l^{0.15}} \tag{27}$$

2. In the rectangular inlet pipe of width b and height h,

$$\zeta_c = 6.05 \left(\frac{Vo}{\upsilon} \cdot \frac{2 \cdot b \cdot h}{b + h}\right) \cdot \frac{(b \cdot h)^{0.94} \cdot D1^{0.12}}{D2^{1.6} \cdot Hi^{0.25} \cdot l^{0.15}} \tag{28}$$

10. Ogawa Experiment

Ogawa derived the equation of pressure drop based upon the idea of the energy dissipation in turbulent vorticity.[17] Namely, energy dissipation Φ of fluid per unit time and per unit volume of fluid can be represented by the following equation

$$\Phi = \eta \cdot |\Omega|^2 \tag{29}$$

where the symbol Ω means a vorticity vector defined as

$$\Omega = \nabla \times \mathbf{V} \tag{30}$$

Here **V** means the fluid velocity vector. Finally Ogawa derived the following equation

$$\zeta_c = \pi \cdot (0.197 \cdot Reb + 2.15) \tag{31}$$

where a symbol Reb is defined as

$$Reb = \frac{|Vrb| \cdot b}{\upsilon_T} \tag{32}$$

υ_T is the eddy kinematic viscosity and the mean value of υ_T in the confined vortex chamber, as shown in Figure 28, is

$$\epsilon = \frac{\upsilon_T}{\upsilon} = 0.05 \cdot Rcy = 0.05 \cdot \left(\frac{Qo}{Hi \cdot \upsilon}\right) \tag{33}$$

In Figure 28, the Kayes and Dial experimental result, and also the Ragsdale experimental result, are shown.

In addition to this, Ogawa's equation was used to estimate the turbulent rotational air flow in the confined vortex chamber and the cyclone by the application of the Burgers' theoretical vortex model. The value of υ_T is not a constant value, but a varied value for the

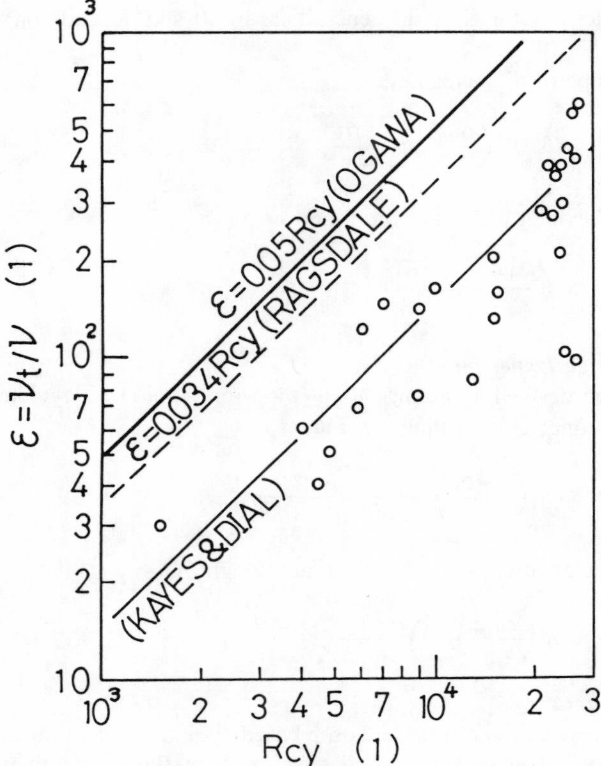

FIGURE 28. Eddy kinematic viscosity for cyclone Reynolds number.

various radial positions in the cyclone. Ogawa and Fujita (1981)[18] measured the value of υ_T in the confined vortex chamber by a hot-wire anemometer and the distributions of υ_T/υ are shown in Figure 17. From this figure, it can be seen that the value of υ_T/υ has the maximum value in the region of the quasi-forced vortex flow and the nearly constant value in the quasi-free vortex flow region.

In order to compare the experimental value of the pressure drop with the empirical equation, Figure 29 shows the detailed construction of cyclone dust collectors.

No. 1 cyclone (D1/D2 = 3, D1 = 120 mm, H_T/D1 = 3.0) was made of celluloid synthetic resin.
No. 2 cyclone (D1/D2 = 3, D1 = 225 mm, H_T/D1 = 3.05) was made of stainless steel.
No. 3 cyclone (D1/D2 = 6, D1 = 300 mm, H_T/D1 = 1.85) was made of zinc plate.

Figure 30 shows the pressure drop Δp_c for the three types of cyclones on the normal room temperature. In the same figure, the three straight lines were theoretically calculated by Equation 31. Further, Figure 31 shows the coefficients of the pressure drops; the straight lines are theoretically calculated values.

From these experimental results of pressure drops, the pressure drop of the cyclone does not directly depend on the large size or small size which is included the miniature cyclone of D1 = 10 to 20 mm. It relates to the intensity of vorticity and the Reynolds stress of the turbulent rotational air flow. Therefore, when the intensity of the vorticity and the Reynolds stress increases to a higher value, the pressure drop increases also.

Figure 32 shows the experimental results of ζc for the three types of cyclones, with the cyclone Reynolds number defined as

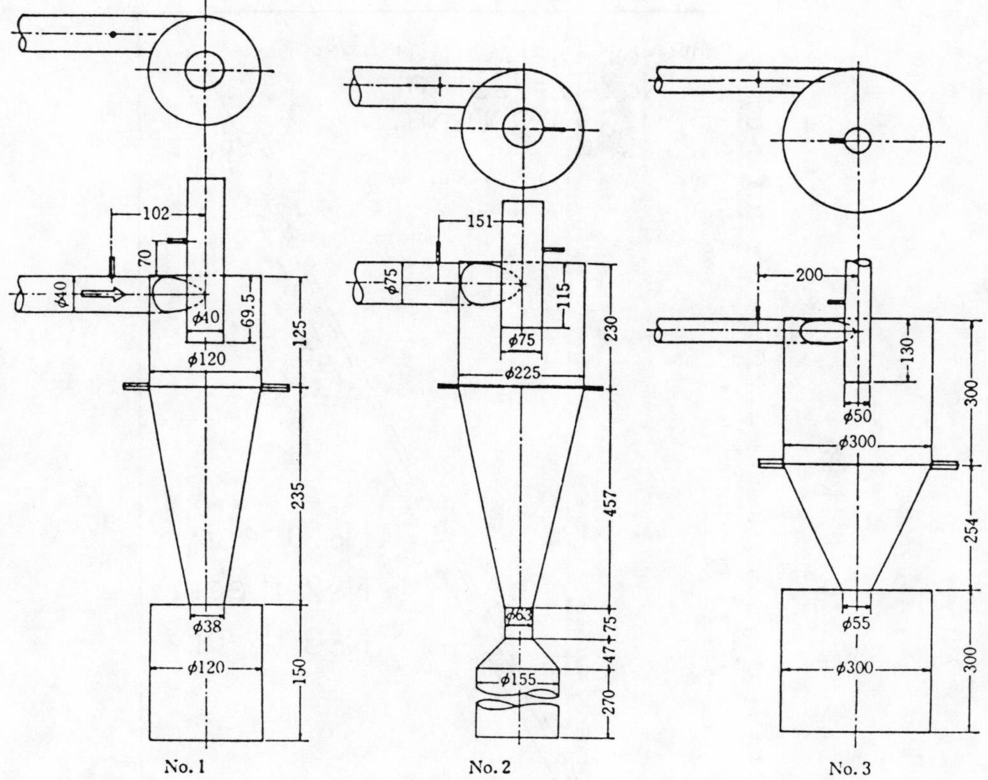

FIGURE 29. Three types of cyclone dust collectors.

$$Rcy = \frac{Qo}{Hi \cdot \upsilon}$$

In the same figure, the lines denoted by A,C,F,H,I,L.S,St are the calculated values of ζc for the No. 2 cyclone (D1 = 225 mm) by the empirical equations as described above. From this figure, we will find that the empirical equations are not always coincidental with the experimental value of ζc. Therefore, if one designs a new type of a cyclone, one should examine the estimation of the value of ζc.

C. Feed-Dust Concentration on the Pressure Drop

When the dust or the fine solid particles are fed with a low feed dust concentration into the cyclone, the pressure drop Δp_c decreases with the dust load in comparison with the pressure drop of the pure air flow. Concerning the influence of the dust-load on the pressure drop, Kriegel (1968)[19] did experiments for the extended feed-dust concentration Co(g/m^3). He used the tangential cyclone with D1/D2 = 2.57 and b/(D1 − D2) = 0.287. Denoting that ζc is the coefficient of the pressure drop for the pure air flow and ζ*c is the dust-laden gas flow, the value of ζ*c/ζc gradually decreases until Co ≒ 2000 g/m^3.

However, when the feed dust concentration Co increases to more than Co = 2000 g/m^3, the value of ζ*c/ζc increases due to the additive pressure drop with the acceleration of the particles, the collisions of particles with each other and with the surface of the cyclone, the friction with particles and the surface of the cyclone, and the float of the particles. This phenomenon is very similar to the characteristics of the pneumatic conveyance of the fine solid particles by pipe.

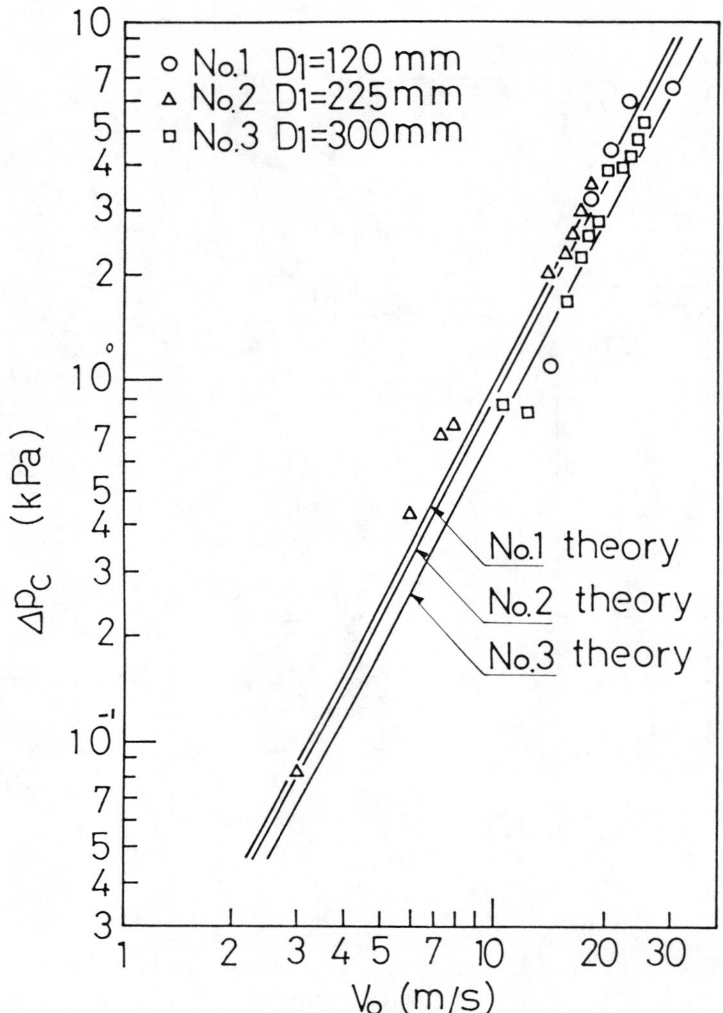

FIGURE 30. Pressure drops for three types of cyclone dust collectors.

On the other hand, Briggs (1946)[20] derived the empirical equation of $\zeta^*c/\zeta c$ as

$$\frac{\zeta^*c}{\zeta_c} = \frac{1}{0.013\sqrt{2.26\cdot Co + 1}} = \frac{77}{\sqrt{2.26\cdot Co + 1}} \tag{34}$$

Ikemori (1955) derived the empirical equation of $\zeta^*c/\zeta c$ as

$$\frac{\zeta^*c}{\zeta_c} = 1 - 0.00367\cdot\sqrt{Co} \tag{35}$$

In addition to this, Barth (1955)[21] did experiments of ζ^*c with wheat on the high mixture ratio

$$m = \frac{\text{(the rate of feed dust quantity, } \dot{G}s\text{(kg/s))}}{\text{(the rate of feed air quantity, } \dot{G}a\text{(kg/s))}}$$

Still more, Troiankin and Balueb (1969)[22] did experiments concerning the pressure drop for the tangential cylindrical cyclone with the several inlet pipes, as shown in Figure 33. Figure 34 shows the experimental results; in this figure e means the surface roughness.

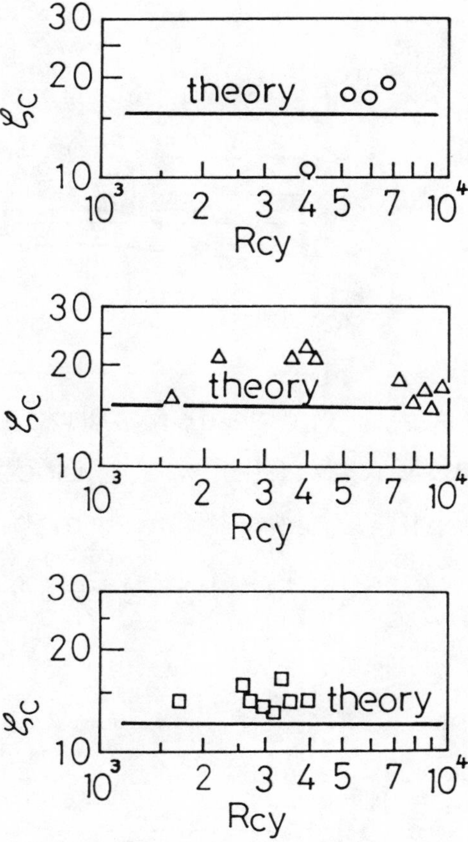

FIGURE 31. Coefficients of the pressure drops for the three types of cyclone dust collectors.

D. Effect of Shapes of the Inner Pipes on Pressure Drops

Figure 35 shows the relative values of pressure drop coefficients for the various types of cyclone inner pipes as measured by Muschelknautz (1980).[23,24]

From an engineering point of view, especially for the maintenance of dust adhering to the inner pipe, the cylinder without auxiliary equipment is best.

V. THEORIES OF CUT-SIZE

A. Rosin, Rammler, and Intelmann Theory

Rosin, Rammler, and Intelmann (1932)[25] derived a theory of cut-size Xc in an ideal separation process for solid particles. They introduced the following assumptions:

1. A drag force acting on a solid particle of a diameter Xp obeys Stokes drag force.
2. The concentration of solid particles in the cross-sectional area of the inlet pipe are distributed homogeneously. A mutual distance between particles is created due to the noninterference of each solid particle's motion.
3. The separated dust never re-entrains.
4. The flow of a dust-laden gas in the cyclone contains the same shape as the inlet pipe of width b and height h. Also this dust-laden gas rotates with the same velocity as the inlet velocity Vo.
5. The rotational dust-laden gas in the cyclone is irrotation (free vortex).

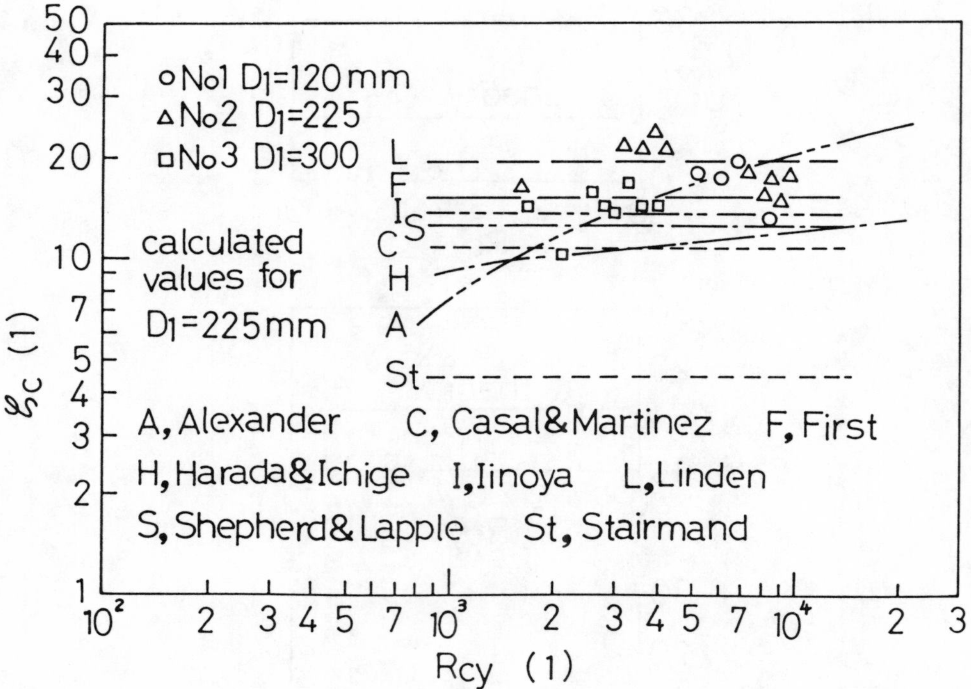

FIGURE 32. Relationship between the coefficient of the pressure drop and the cyclone Reynolds number.

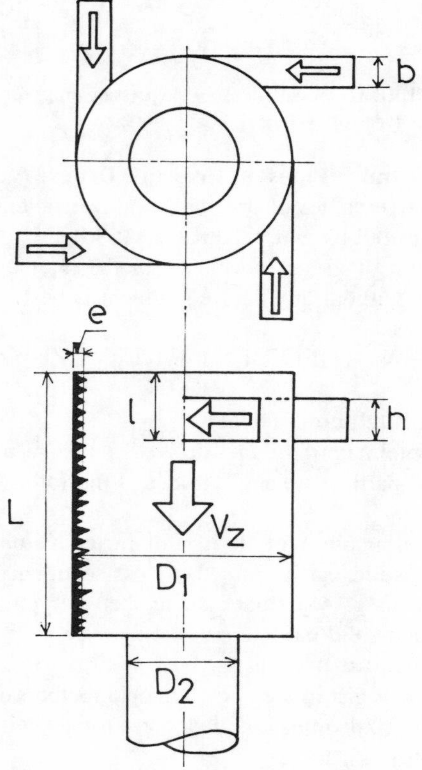

FIGURE 33. Illustration of the cylindrical cyclone dust collector.

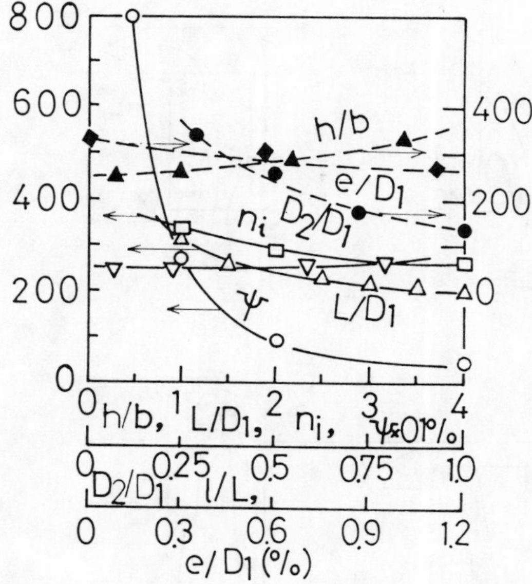

D1 (mm)	D2/D1 (1)	n_i	e/D1 (1)	ψ (1)	h/b (1)	L/D1 (1)
470	0.530	2	5×10^{-3}	7.4	1	1.25
440	0.625	4	6×10^{-3}	5.6	1	1.15
610	0.525	1:2:4	5×10^{-3}	2.5:4.5:5.9	1	2.34
1600	0.262	4	2.5×10^{-3}	8.6	1.69	1.50

FIGURE 34. Experimental results of cylindrical cyclone dust collector of Figure 33.

They assumed that the centrifugal force acting on a solid particle is equal to the drag force described as

$$\rho_p \cdot \frac{\pi \cdot Xp^3}{6} \cdot \frac{Vo^2}{r} = 3 \cdot \pi \cdot \eta \cdot Xp \cdot \left(\frac{dr}{dt}\right) \tag{36}$$

Using an initial condition of a solid particle of the radius r = (D1/2) − S at a time t = 0, the following equation can be obtained as

$$r^2 = \frac{\rho_p \cdot Xp^2 \cdot Vo^2}{9 \cdot \eta} \cdot t + \left(\frac{D1}{2} - S\right)^2 \tag{37}$$

Then a lapse of time t of the moving particle may be described nearly by the number N of rotation of the solid particle in the cyclone as

$$t = \frac{\pi \cdot D1}{Vo} \cdot N \tag{38}$$

At the same time t, the solid particle reaches from (D1/2) − S to the outer wall (r = D1/2). Substituting Equation 38 into Equation 37, we can obtain the equation of the particle size Xs corresponding to the above described boundary condition as

$$Xs = 3\sqrt{\frac{S \cdot (D1 - S) \cdot \eta}{\pi \cdot \rho_p \cdot D1 \cdot N \cdot Vo}} \tag{39}$$

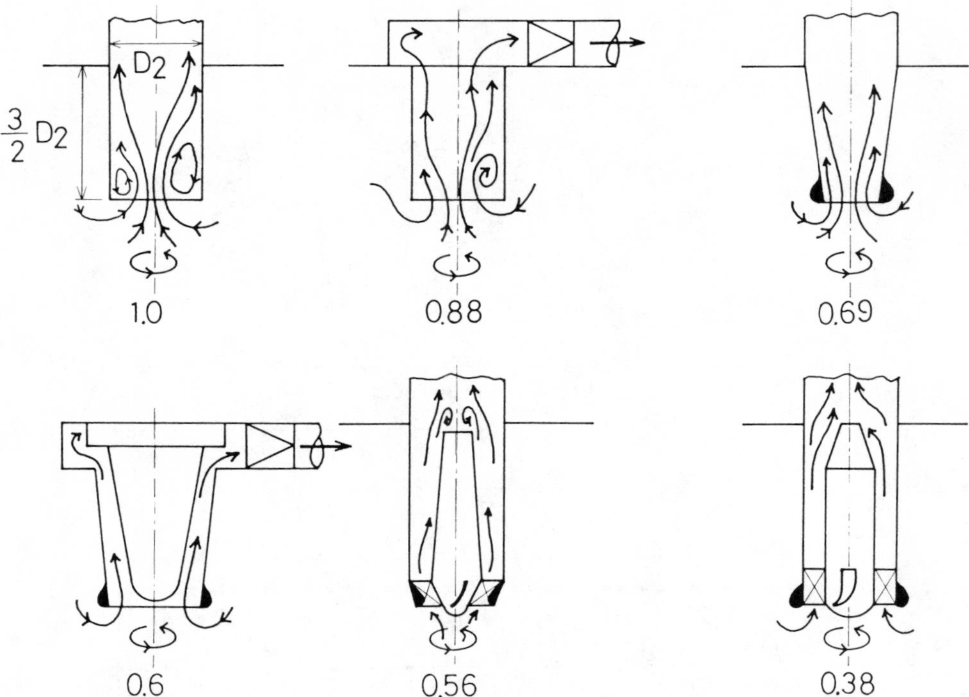

FIGURE 35. Relative values of the coefficients of the pressure drops for various types of inner pipes.

Therefore, the minimum separate particle of size Xsc may be regarded as a particle that is introduced into the cyclone at the position S = b, as shown in Figure 36. Finally, Equation 39 becomes

$$Xsc = 3\sqrt{\frac{b\cdot(D1 - b)\cdot\eta}{\pi\cdot\rho_p\cdot D1\cdot N\cdot Vo}} \tag{40}$$

In this figure, the calculated example of Xsc for D1 = 150 mm, b = 50 mm, ρ_p = 2 g/cm^3, η = 1.8 × 10^{-5} Pa·s, and N = 1 is shown.

B. Davies' Theory

Davies (1952)[26] derived a formula of the cut-size Xmin based upon the assumption of a relationship between the transverse distance (R1 − R2) of the solid particle in the free vortex flow and the lapse of time δt = Ht/Vo of fluid flow for the downward direction in the cyclone as shown in Figure 37. Then the minimum separate particle size Xmin can be obtained as follows:

$$Xmin = \sqrt{\frac{9\cdot\eta\cdot D1^2}{8\cdot(\rho_p - \rho)\cdot Vo\cdot Ht}\left\{1-\left(\frac{D2}{D1}\right)^2\right\}} \tag{41}$$

The calculated example of Equation 41 for D1/D2 = 3, 2Ht/D1 = 4, ρ_p = 2 g/cm^3, and η = 1.81 × 10^{-5} Pa·s is shown for D1 = 0.075, 0.15, 0.3, and 0.6 m in Figure 37.

C. Goldshtik Theory

Goldshtik (1962)[27] derived the equation of the cut-size Xmin under the following assumptions:

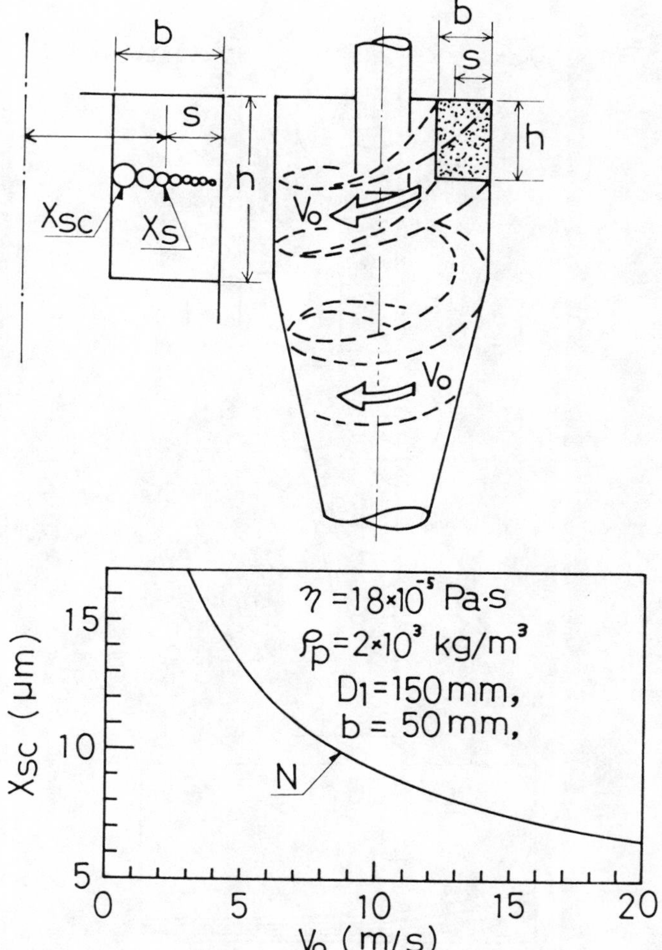

FIGURE 36. Illustration of the cut-size of Rosin, Rammler, and Intelmann (1932).

1. The drag force acting on a solid particle obeys Stokes drag force.
2. The inertia force of a solid particle is negligible in comparison with the centrifugal force acting on a solid particle.
3. The motion of a fine solid particle nearly follows the motion of the fluid flow for the axial and tangential directions.

In addition to this, he considered the mechanically stable and unstable motions of the fine solid particle for the separation process in the cyclone. Finally the minimum cut-size X_T.min for 100% separation can be written as

$$X_T\cdot\text{min} = \sqrt{\frac{9}{2\,\pi}\cdot\frac{\eta}{\rho_p\cdot L}\cdot\frac{Ao}{\lambda^2}\left\{1+\left(\frac{D2}{D1}\right)^2\right\}\frac{1}{Vo}} \qquad (42)$$

where the symbol λ is a dimensionless number defined as $V\theta\cdot r = \lambda\cdot Vo\cdot Ro$.

Then the calculated example of X_T.min for R1 = 75 mm, R2 = 25 mm, L = 500 mm, Do = 50 mm, ρ_p = 2 g/cm^3, and $\eta = 1.8 \times 10^{-5}$ Pa·s is shown for λ = 0.8 and 1.0 in Figure 38.

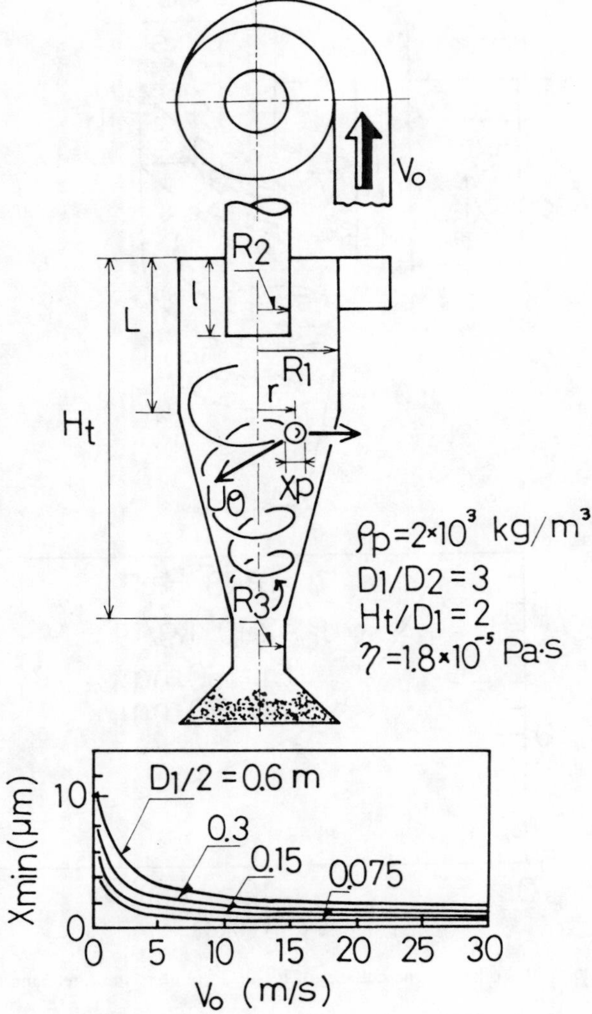

FIGURE 37. Illustration of the cut-size of Davies (1952).

D. Fuchs' Theory

1. Theory of Cut-Size

In order to derive the equation of cut-size, Fuchs (1964)[28] assumed that the tangential velocity Vθ of fluid flow near the outer wall could be written

$$V\theta = \frac{Vo}{2} \cdot \sqrt{\frac{R1}{r}} \tag{43}$$

Then the minimum cut-size Xmin for the 100% separation possibility can be written

$$Xmin = \sqrt{\frac{6 \cdot \eta \cdot b}{\pi \cdot \rho_p \cdot N \cdot Vo}} \tag{44}$$

The calculated example of Xmin for b = 50 mm, ρ_p = 2 g/cm^3, N = 1, η = 1.8 × 10^{-5} Pa·s is shown in Figure 39 and in Figure 40 as a dotted line. In addition to this, Fuchs' idea for the total collection efficiency is shown in this figure.

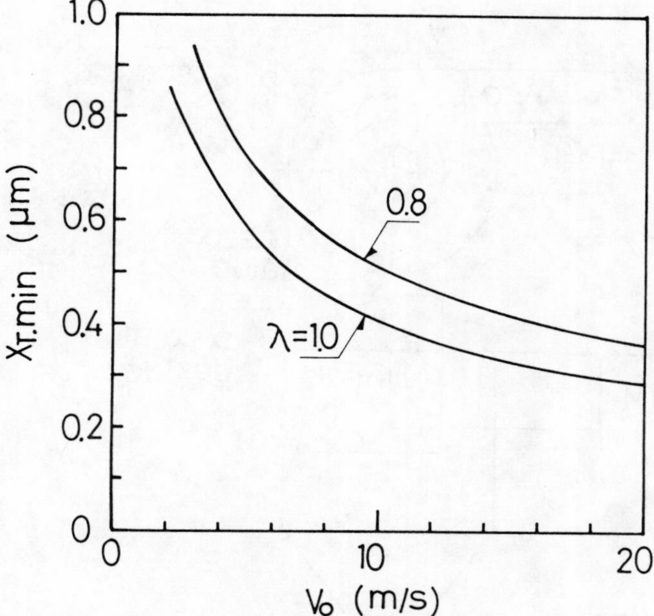

FIGURE 38. Calculated examples of the cut-size of Goldshtik (1962).

2. *Collection Efficiency in the Turbulent Rotational Flow*

We regard the motion of solid particles in the cyclone as a state of continuous turbulent mixing. The tangential velocity Vθ of the fluid near the outer wall is nearly equal to Vθ/2, as shown in Figure 40. Then the radial velocity Ur of the solid particles can be written

$$Ur = \frac{\tau \cdot Vo^2}{4 \cdot R1} \tag{45}$$

where a symbol τ is $\rho_p \cdot Xp^2/18 \cdot \eta$.

Here, denoting that the symbol n is the particle number per unit volume of space, the particle number I by the centrifugal sedimentation on the outer wall surface per unit time can be represented as follows:

$$I = \frac{\tau \cdot Vo^2}{4 \cdot R1} \cdot n \tag{46}$$

Then denoting that L is the height of the cyclone and N is the revolution number of the particle as shown in Figure 41, the value of I·L/N means the sedimentation particle number per unit time on the outer wall surface.

The total particle number per unit time into the cyclone can be written as b·h·n·Vo. Therefore, the entering particle number at the place z becomes nearly b·h·n·Vo/2, and also the escaping total particle number at the place z + dz and at the small area h dz nearly becomes

$$b \cdot h \cdot \frac{Vo}{2}\left(n + \frac{dn}{dz} \cdot dz\right) + h \cdot dz \cdot \frac{\tau \cdot Vo^2}{4 \cdot R1} \cdot n \tag{47}$$

Consequently a differential equation for the remaining particles in the small volume can be written

$$-\frac{1}{n} \cdot \frac{dn}{dz} = \frac{\tau \cdot Vo}{2 \cdot R1 \cdot b} \tag{48}$$

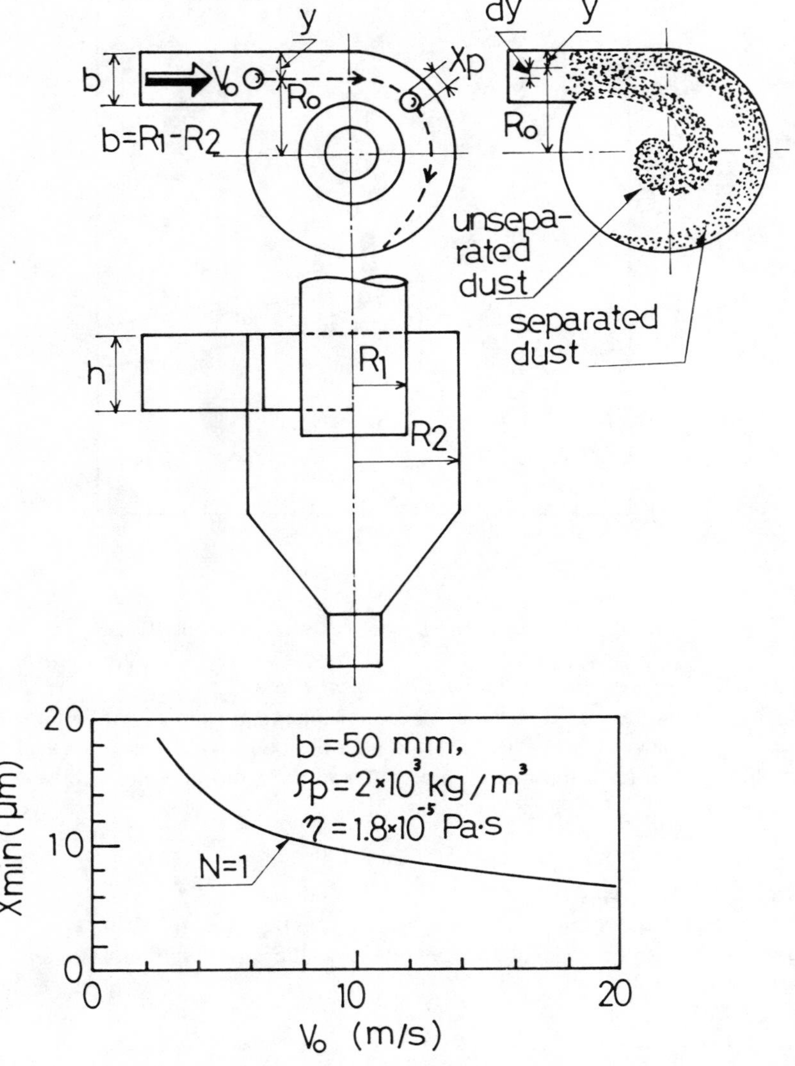

FIGURE 39. Illustration of the cut-size of Fuchs (1964).

Here, denoting that z is a special length z = $2 \cdot \pi \cdot R1 \cdot N$ and n_o is the particle concentration at z = 0, we can obtain the following equation

$$\frac{n}{n_0} = \exp\left(-\frac{\pi \cdot Vo \cdot \tau}{b} \cdot N\right) \tag{49}$$

Then, the fractional collection efficiency $\eta_x(Xp)$ for the particle size Xp can be represented as

$$\eta_x(Xp) = \frac{n_0 - n}{n_0} = 1 - \exp\left(-\frac{\pi \cdot \rho_p \cdot Vo \cdot N}{18 \cdot \eta \cdot b} \cdot Xp^2\right) \tag{50}$$

From this equation, the calculated example of the cut-size Xc corresponding to $\eta_x(Xp)$ = 0.5 is shown by the bold line for b = 5 cm, ρ_p = 2 g/cm³, $\eta = 1.8 \times 10^{-5}$ Pa·s, and N = 1 in Figure 40.

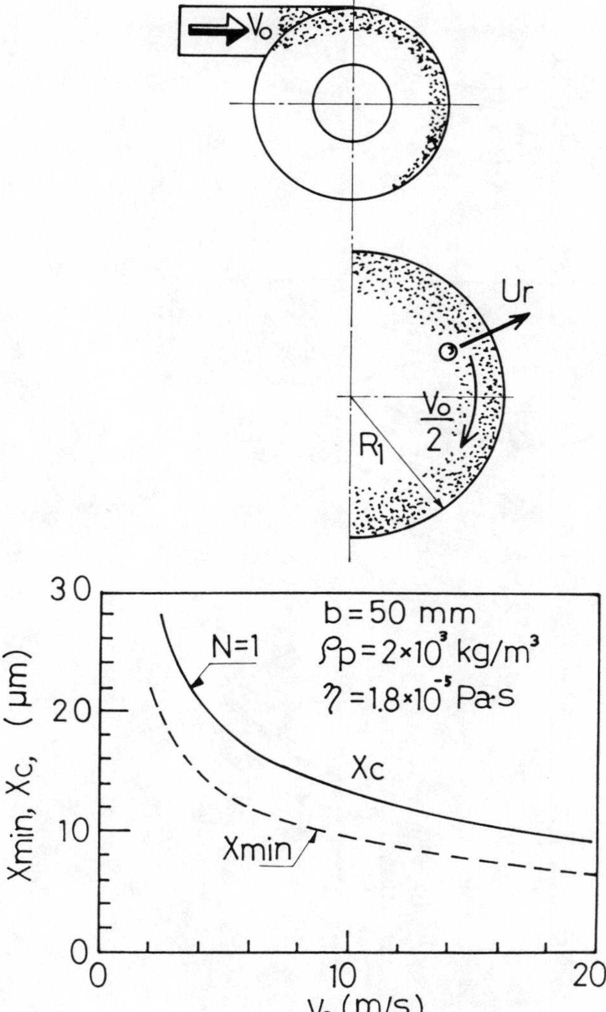

FIGURE 40. Cut-size for the turbulent rotational flow by Fuchs.

E. Ogawa's Theory

According to the same physical idea of Barth, Ogawa (1977)[29] derived a theoretical formula of the cut-size Xc. Assuming that the centrifugal force acting on a solid particle of an arbitrary radius r is equal to the Stokes drag force by the inward radial air flow (Figure 42), the mechanical equilibrium particle rotates steadily with the velocity Uθ on an arbitrary radius r. This state of mechanical equilibrium can be written

$$\rho_p \frac{\pi \cdot Xp^3}{6} \cdot \frac{U\theta^2}{r} = -\phi \cdot 3\pi \cdot \eta \cdot Xp \cdot Vr \tag{51}$$

where ϕ means a shape factor of a solid particle of the diameter Xp. We also regard that the tangential velocity Uθ of the solid particle is nearly equal to that (Vθ) of the fluid. Then the diameter of this mechanical equilibrium particle on the radius r denotes by Xp = X_B, so Equation 51 transforms to

$$X_B = \sqrt{\frac{\phi \cdot 18 \cdot \eta \cdot (-Vr) \cdot r}{\rho_p \cdot V\theta^2}} \tag{52}$$

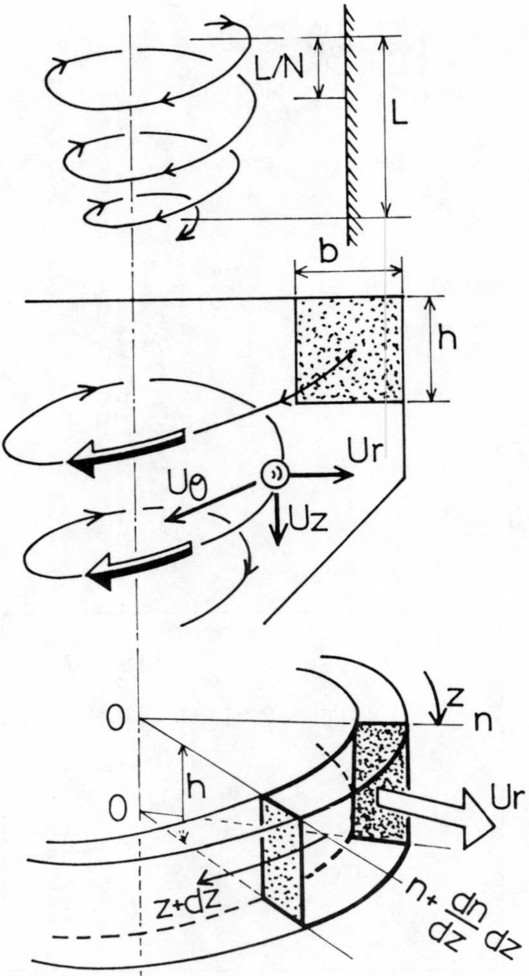

FIGURE 41. Illustration of the collection efficiency on the turbulent rotational flow by Fuchs.

The cut-size Xc can be defined as a mechanical equilibrium particle on a radius r = a of the imaginary cylindrical surface. Also the particles of the cut-size Xc are separated 50% on the imaginary cylindrical surface and the remaining 50% escaped from the inner pipe.

Therefore, Equation 52 becomes

$$Xc = \sqrt{\frac{18 \cdot \phi \cdot a \cdot (-Vra) \cdot \eta}{\rho_p \cdot V\theta a^2}} \tag{53}$$

where Vθa and Vra are the tangential and radial velocities of fluid on the radius r = a.

Here, the flow rate Qa in the cyclone may be written as

$$Qa = Qo\left(1 - \frac{Qb}{Qo}\right) \tag{54}$$

where Qb is the flow rate into the dust bunker directly from the inlet pipe. The inward radial velocity −Vra in the quasi-forced vortex of Ogawa's combined vortex flow can be written

$$Vra = -\frac{a \cdot Qo\left\{1 - (Qb/Qo)\right\}}{2 \cdot \pi \cdot \dot{m} \cdot Ht \cdot r_t^2} \tag{55}$$

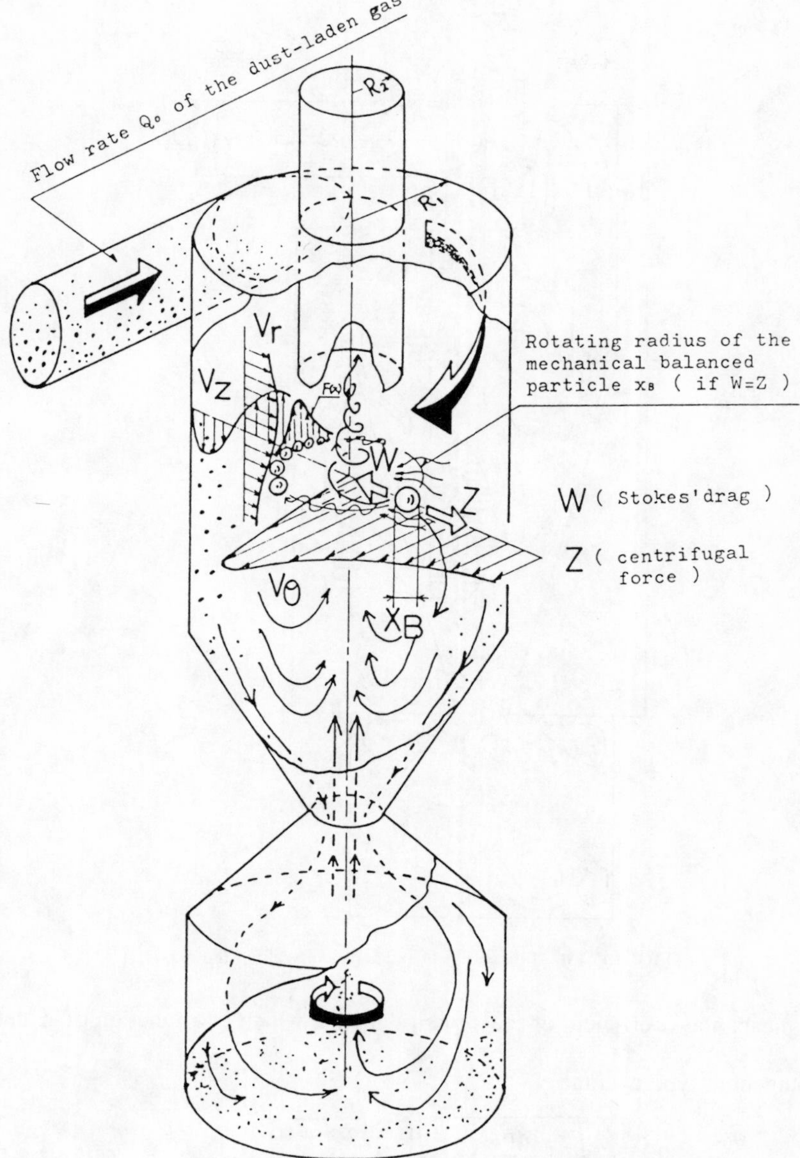

FIGURE 42. Illustration of the mechanical balanced particle by Ogawa (1977).

where m′·Ht is the effective imaginary cylindrical height, as shown in Figure 43. Also the measured distributions of the radial velocity Vr for Vo = 5, 10, and 15 m/s are shown for D1/D2 = 3 and Do = 50 mm in Figure 44. In addition, the empirical equation of the flow rate Qb into the dust-bunker can be written

$$\frac{Qb}{Qo} = 2.62\left\{1 - 0.36\left(\frac{D3}{D1}\right)^{-0.56}\right\}\left(\frac{D3}{D1}\right)^{1.16} \tag{56}$$

Next, when the fluid with the velocity Vo in the inlet pipe enters the cyclone, the tangential velocity Vó near the outer wall may be written

$$V\acute{o} = k \cdot Vo \tag{57}$$

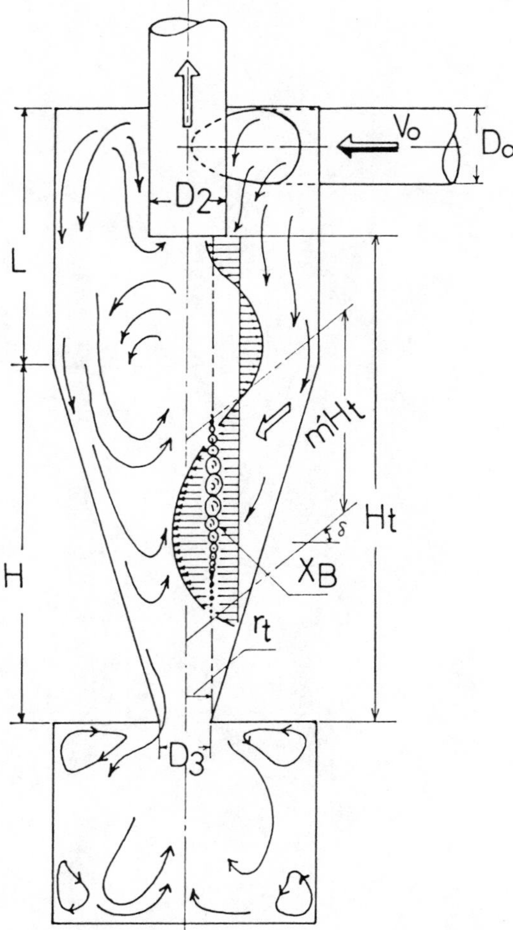

FIGURE 43. Illustration of radial velocity along the z-axis.

where k means the coefficient of the velocity change which is a function of Λ defined as:

for the tangential type cyclone

$$\Lambda = 0.332 \sqrt{\left(1 - \frac{Do}{D1}\right) \cdot \frac{H_T}{R1}} \sqrt{\frac{2 \cdot \pi \cdot \nu \cdot H_T}{Ao \cdot Vo}} \cdot \frac{D1}{\sqrt{Ao}} \tag{58}$$

for the Linden type cyclone

$$\Lambda = 0.332 \sqrt{\left(1 + \frac{b}{D1}\right) \cdot \frac{H_T}{R1}} \sqrt{\frac{2 \cdot \pi \cdot \upsilon \cdot H_T}{Ao \cdot Vo}} \cdot \frac{D1}{\sqrt{Ao}} \tag{59}$$

and the value of k is shown in Figure 45.

Therefore, the equation of the quasi-free vortex can be written

$$\Gamma n = k \cdot Vo \left(R1 - \frac{Do}{2}\right)^n \tag{60}$$

where a relationship between a, n, and r_t can be defined as

$$\frac{2 \cdot a}{r_t} = \frac{2 + n}{1 + n} \tag{61}$$

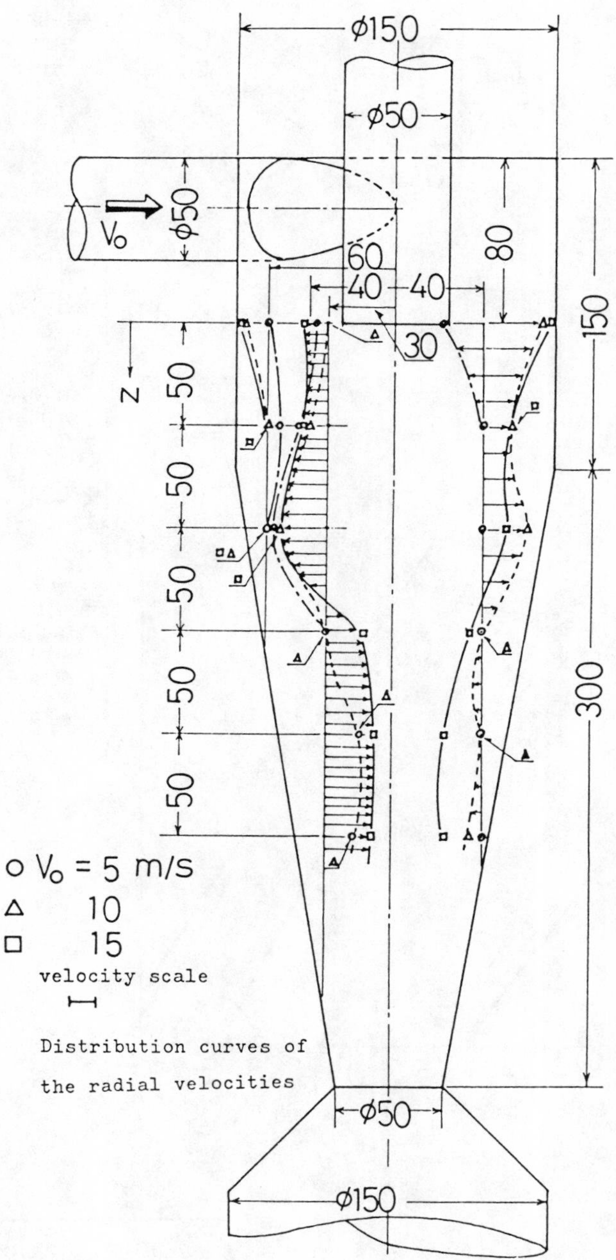

FIGURE 44. Distributions of the radial velocity of air along the z-axis.

So the equation of k becomes

$$k = \frac{I\,n\cdot(2+n)^2}{2\cdot a\cdot(1+n)}\left\{\frac{2+n}{2\cdot a\cdot(1+n)}\right\}^n \tag{62}$$

Consequently, substituting Equations 54, 55, and 61 into Equation 53, we can obtain the formula of the cut-size Xc defined as

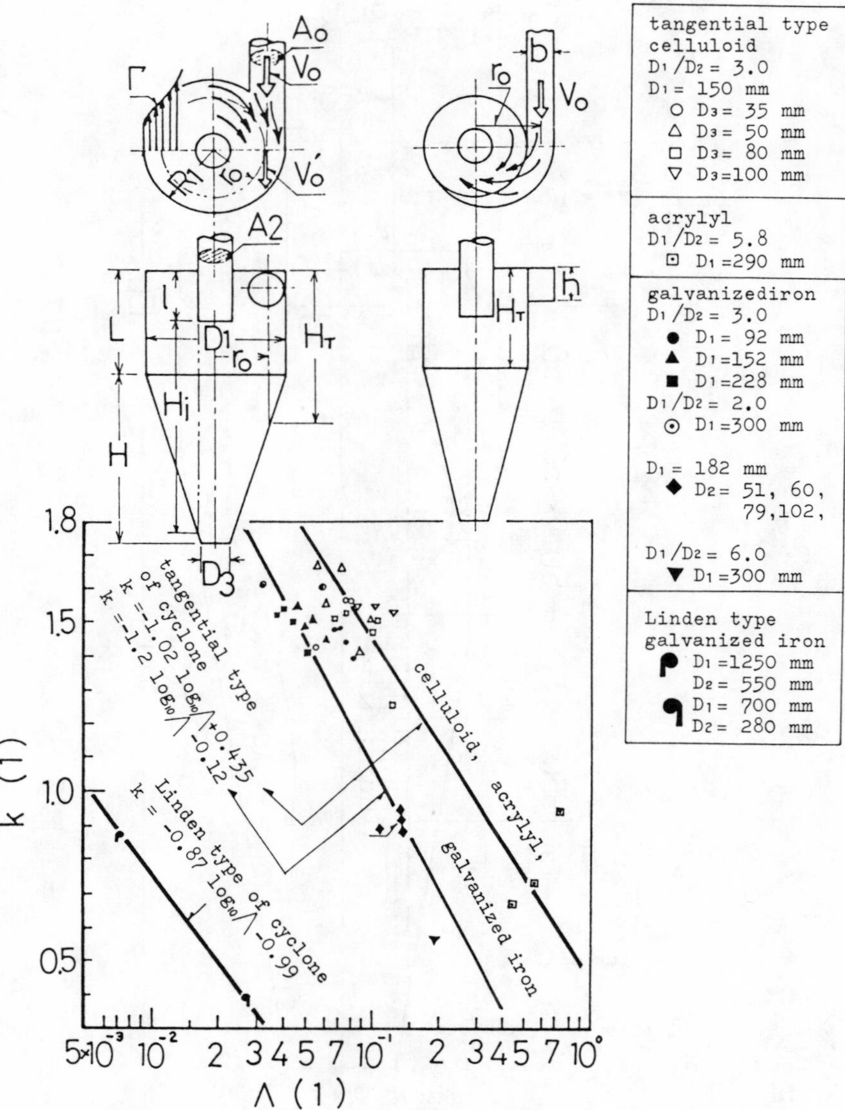

FIGURE 45. Coefficient of the velocity change.

$$Xc = \frac{2}{1+n} \cdot \left(\frac{1+n}{2+n}\right)^n \left\{\frac{4\,D2}{3 \cdot D1\left(1 - \frac{Do}{D1}\right)}\right\}^n \sqrt{\frac{\phi \cdot 9 \cdot \eta \cdot Ao\left\{1 - (Qb/Qo)\right\}}{\pi \cdot m' \cdot k^2 \cdot \rho_p \cdot Ht \cdot Vo}} \tag{63}$$

From this equation, we may say that even if the diameter D1 of the cyclone decreases to a small type, it is not always possible to decrease the cut-size Xc. This fact is very important in designing the cyclone.

F. Barth's Theory

1. Theory of Cut-Size

As shown in Figure 46, Barth (1954)[30,31] derived the equation of cut-size which corresponds to the particle diameter of a state of mechanical equilibrium between the centrifugal force

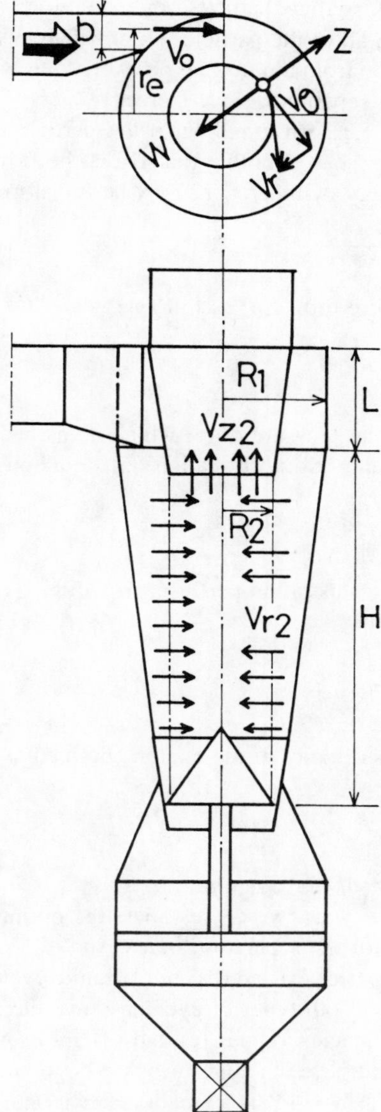

FIGURE 46. Illustration of the cut-size of Barth (1954).

Z by the tangential velocity Vθ2 and the Stokes drag force W by the inward radial velocity Vr2 on a radius R2 of the imaginary cylindrical surface.

Now this state of mechanical equilibrium can be written

$$\rho p \frac{\pi \cdot Xp^3}{6} \cdot \frac{V\theta 2^2}{R2} = 3 \cdot \pi \cdot \eta \cdot Xp \cdot Vr2 \tag{64}$$

Here, defining the terminal velocity Wsc as

$$Wsc = \frac{\rho p \cdot Xp^2 \cdot g}{18 \cdot \eta} \tag{65}$$

Equation 64 is transformed to

$$Wsc = \frac{Vr2 \cdot R2 \cdot g}{V\theta 2^2} \tag{66}$$

Therefore, all of the particles swifter than Wsc on a radius r = R2 cannot enter into the imaginary cylindrical surface; all of the particles slower than Wsc on a radius r = R2 flow through this surface and escape from the inner pipe. Accordingly, this imaginary cylindrical surface may be regarded as a separation sieve (screen).

Barth defined the cut-size Xc corresponding to the particle diameter having a terminal velocity Wsc of Equation 66. Then denoting that Qo is the flow rate into the cyclone and H is the height of the imaginary cylinder, Vr2 may be transformed to

$$Vr2 = \frac{Qo}{2 \cdot \pi \cdot R2 \cdot H} \qquad (67)$$

Further, he defined the pressure drop Δp_c as follows:

$$\Delta p_c = \epsilon \cdot \rho \cdot \frac{V\theta 2^2}{2} \qquad (68)$$

where ϵ is the coefficient of the pressure drop, which he named "aerodynamische Güte".[3]

Substituting Equations 67 and 68 into Equation 66, Equation 66 is transformed into

$$Wsc = \frac{\epsilon \cdot \rho \cdot g \cdot Qo}{4 \cdot \pi \cdot H \cdot \Delta p_c} \qquad (69)$$

Now in order to represent the separation performance of the cyclone, Barth introduced the separation index Ab defined as

$$Ab = \frac{Wsc \cdot \overline{V}z1}{g \cdot D1} \qquad (70)$$

where $\overline{V}z1$ is the mean axial velocity in the cyclone defined as

$$\overline{V}z1 = \frac{4 \cdot Qo}{\pi \cdot D1^2}$$

2. Performance Comparison of the Cyclones

In the selection of cyclone types, we can estimate the optimum cyclone type for a fixed space by applying the value of the separation index Ab.

As shown in Figure 47, types of (a) and (b) are hydraulic cyclones, but they can be treated the same as the gas cyclones. Both types of cyclones have the small quantities of flow rate and the high separation efficiencies. Types (c) and (d) are examples of large size cyclones and are applied to the blast furnace. In spite of being large-size cyclones, they display the high separation performance. Type (e) is a small-size cyclone.

Types (f) and (g) are used for the huge flow rate of gas and at the same time for the narrow set-up area. These types of constructions are applied to multi-cyclones. Type (h) is an axial cyclone of reversed flow. When the relationship between the cross-sectional area of the inlet pipe (D1) and that of the outer pipe (D2) is satisfied as

$$\frac{\pi \cdot D1^2}{8} = \frac{\pi \cdot D2^2}{4}$$

so the pressure drop Δp_c of this cyclone becomes

$$\Delta p_c = 2 \cdot \rho \cdot \frac{Vz2^2}{2}$$

Then the coefficient §d of the pressure drop (Δp_c = §d·ρ·$\overline{V}z^2$1/2) becomes 16. Type (i) is used for the case of Ab → 0.

Figure 48 shows a relationship between the value of §d and the value of Ab. When the value of Ab becomes a large value, the effect of the value of Vθ2/Vz2 may be neglected.

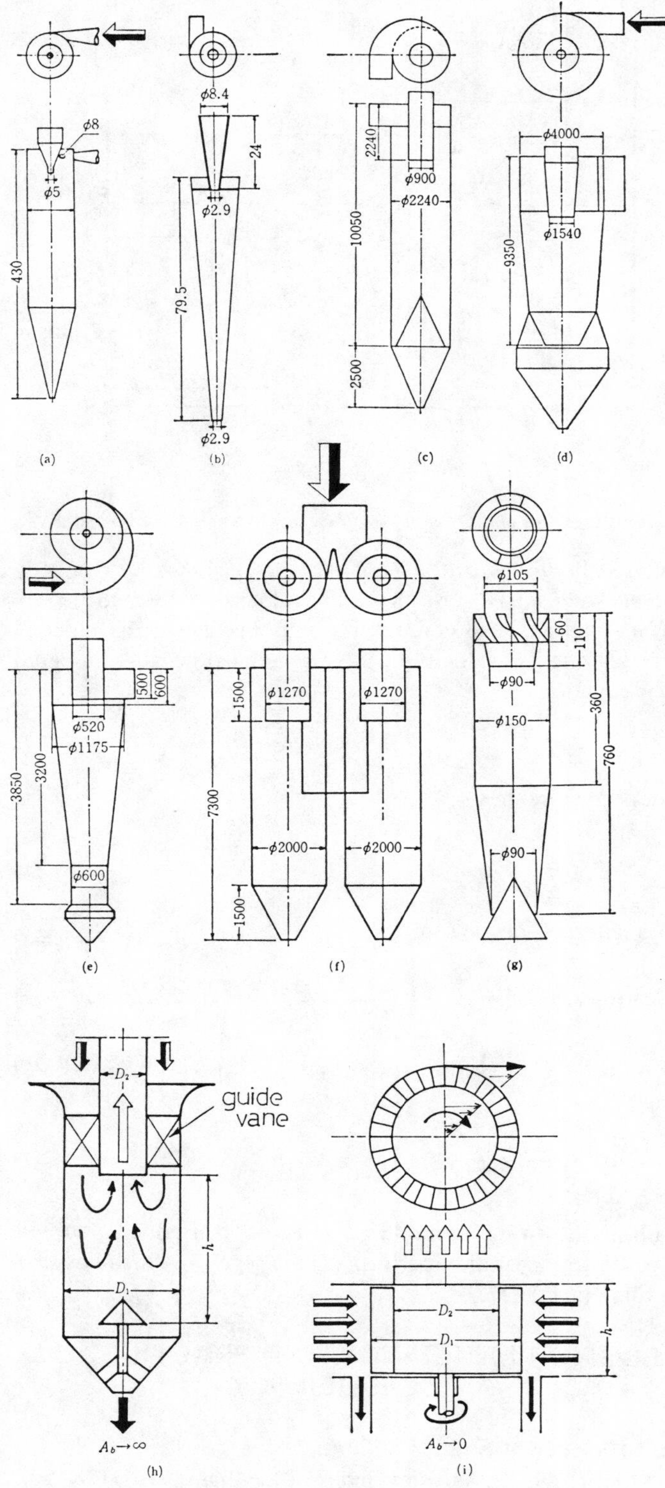

FIGURE 47. Several types of cyclone dust collectors.

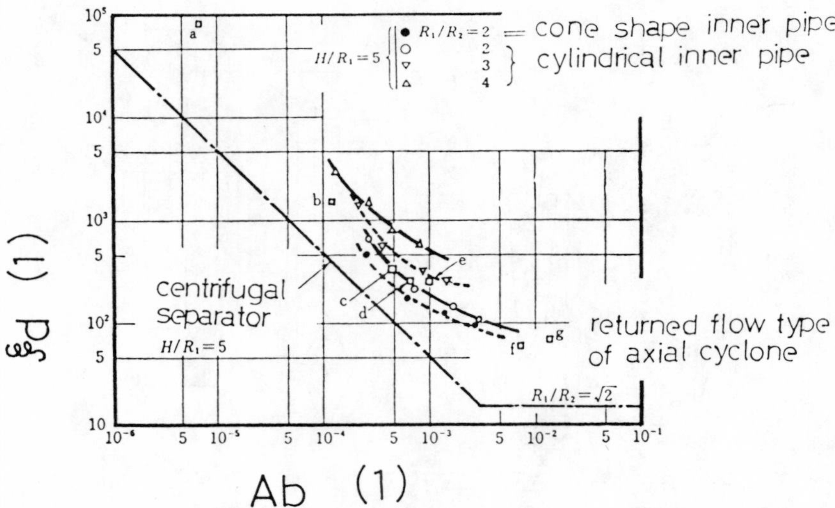

FIGURE 48. Relationship between the coefficient of the pressure drop and the separation index.

This is the reversed flow type of the axial cyclone as shown by type (h). On the contrary, when the value of Ab becomes zero, the radial velocity Vr2 and the axial velocity Vz2 may be designed to a small value in comparison with the tangential velocity. Therefore in the case of the constant angular velocity of the rotational flow, the pressure drop Δp_c is equal to the following equation

$$\Delta p_c = \rho \cdot \frac{V\theta 1^2}{2}$$

Then the value of §d becomes

$$§d = \left(\frac{V\theta 1}{\bar{V}z1}\right)^2 \tag{71}$$

In this case, Equation 69 becomes

$$Wsc = \frac{g \cdot Qo}{2 \cdot \pi \cdot H \cdot V\theta_1^2} = \frac{\pi \cdot R1^2 \cdot \bar{V}z1 \cdot g}{2 \cdot \pi \cdot H \cdot V\theta_1^2} = \frac{g \cdot R1^2}{2 \cdot \pi \cdot \bar{V}z1 \cdot §d} \tag{72}$$

From this equation, a relationship between §d and Ab (= Wsc·$\bar{V}z^1/_2$·R1·g = R1/4·H·§d) becomes

$$§d = \frac{1}{4 \cdot Ab(H/R1)} \tag{73}$$

This relationship is shown in Figure 48 for H/R1 = 5. Further, you will find that the best type of the cyclone for separation performance is type (a) which has the small value of Ab and the large value of D1/D2.

VI. COLLECTION EFFICIENCY AND FRACTIONAL COLLECTION EFFICIENCY

A. Collection Efficiency of Small Cyclones

As shown in Figure 49, the small cyclones of the diameters D1 = 50, 75, 100, 125, and 150 mm, the total collection efficiencies $\eta_c(\%)$ with the test dust of Kanto-Loam (ρ_p = 2.97 g/cm^3, X_{R50} = 9.0 μm), as shown in Figure 50, are shown for the mean inlet velocity

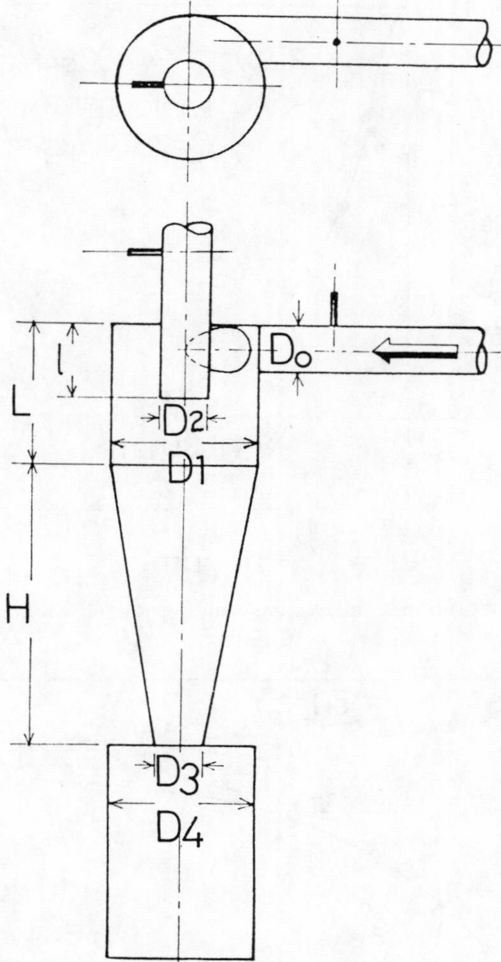

main dimensions

symbols	C-50 (mm)	C-75 (mm)	C-100 (mm)	C-125 (mm)	C-150 (mm)
D_0	17	25	33	41.7	50
D_1	50	75	100	125	150
D_2	17	25	33	41.7	50
D_3	17	25	33	41.7	50
D_4	50	75	100	125	150
L	50	75	100	125	150
l	27	38	52.8	66.7	80
H	100	150	200	250	300

FIGURE 49. Similarity types of cyclones.

Vo in Figure 51. From this figure, it can be seen that the total collection efficiency η_c is increased with intensifying the inlet velocity Vo until Vo = 20 m/s for the cyclone diameter D1 = 150 mm, and that η_c is decreased with increasing the inlet velocity Vo for the cyclone diameter D1 = 50, 75, 100, and 125 mm, respectively.

When the inlet velocity Vo becomes Vo = 20 to 30 m/s, the re-entrainment of the separated dust, the diffusion of fine solid particles by the turbulent velocity, and the repulsion

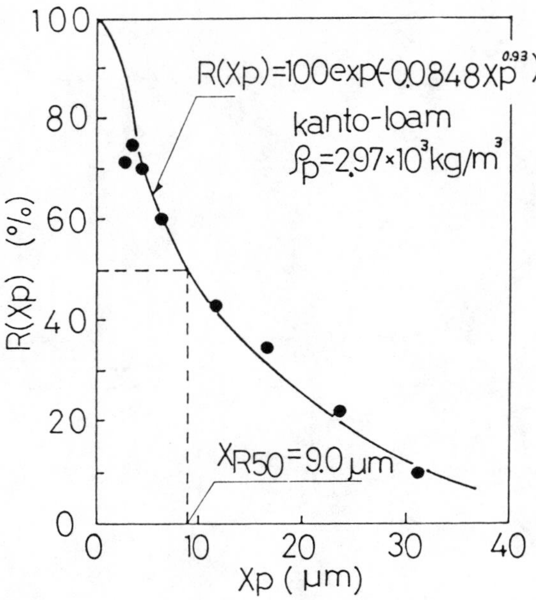

FIGURE 50. Residue distribution of Kanto-Loam as a test dust.

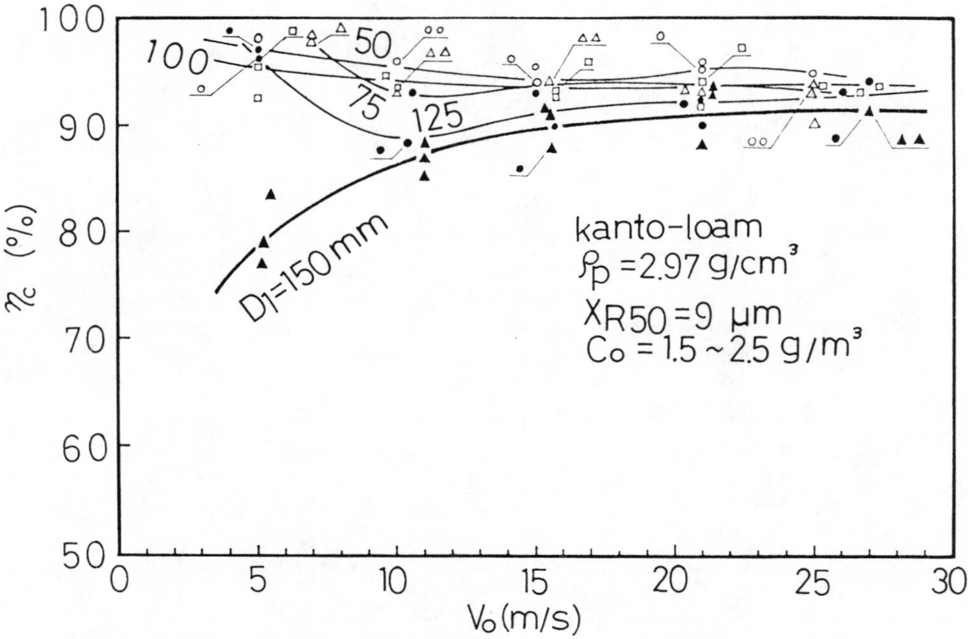

FIGURE 51. Total collection efficiency of similar types of cyclones for inlet velocity.

of the solid particles on the cyclone outer-wall occur. Therefore, the total collection efficiency begins to decrease with increasing the inlet velocity Vo. This phenomenon becomes remarkable for the cyclone diameters D1 = 100, 75, and 50 mm, respectively.

From the practical point of view, diameter D1 of the cyclones should not be decreased smaller than D1 = 100 mm. As described in the Ogawa theory of the cut-size Xc, the cut-size is not always decreased by decreasing the diameter of the cyclone form D1 = 100 mm to D1 = 50 mm. This fact is very important for designing new types of cyclones.

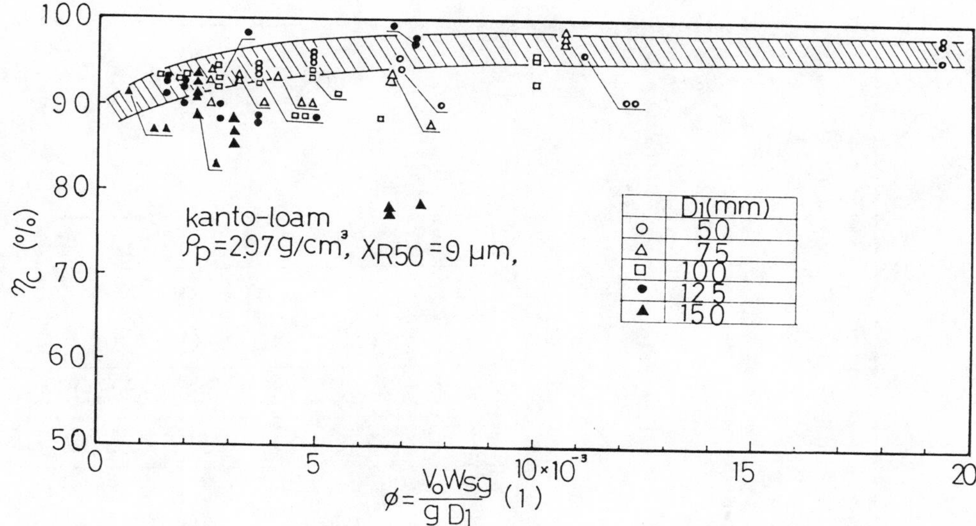

FIGURE 52. Collection efficiency for similar types of cyclones for the inertia parameter.

On the other hand, the collection efficiency η_c of Figure 51 is rewritten by the dimensionless inertia parameter ϕ

$$\phi = \frac{Vo \cdot Wsg}{g \cdot D1}$$

as shown in Figure 52, where Wsg is the terminal velocity of the Kanto-Loam which is based upon the cut-size estimated from the Ogawa theory. The feed dust-concentration is limited between Co = 1.5 to 2.55 g/m³.

B. Fractional Collection Efficiency

Figure 53 shows the curves of the fractional collection efficiencies for the cyclone of D1 = 150 mm of Figure 49. The inlet velocities Vo are 4.94, 10.3, 15.1, and 19.7 m/s, respectively. Those curves were determined by the method of Schmidt. Test dust was Kanto-Loam of Figure 50. The feed concentration Co of dust was Co = 1.5 to 2.5 g/m³. Then, from this figure, the fractional collection efficiency $\eta_x(Xp)$ can be described by

$$\eta_x(Xp) = 1 - \exp\left\{-\alpha \cdot \left(\frac{Xp}{Xc50}\right)^m\right\} \qquad (74)$$

The separation index m decreases with decreasing the cut-size and the value of α is α = 0.693 (= ln2).

The experimentally determined cut-sizes Xc50 which correspond to $\eta_x(Xp)$ = 0.5 (50%) are written in this figure. Further, the comparison of the experimentally determined cut-size Xc50 with the theoretically estimated cut-size Xc of Ogawa's theory is shown in Figure 54.

VII. A NEW IDEA FOR INCREASING COLLECTION EFFICIENCY BY ABRAHAMSON AND LIM

Abrahamson and Lim (1981)[32,33] developed a new idea for the improvement of the collection efficiency η_c by modifications to the air flow within the exhaust pipe. For example, just within the exhaust lip, the inward radial velocity of air dropped to zero, or even reversed, while the tangential velocity of air did not diminish, so that feed dust which could not be

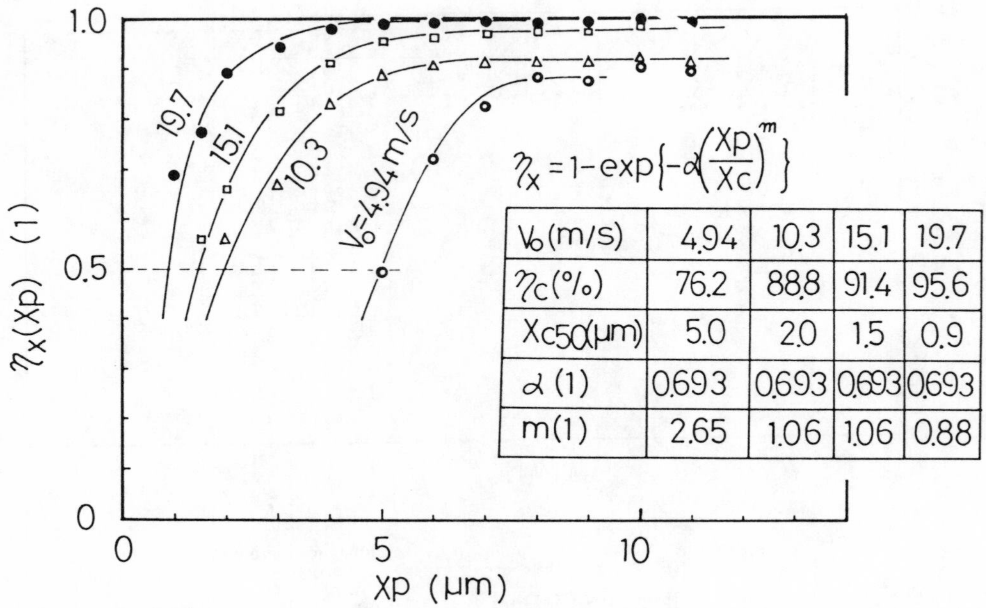

Vo(m/s)	4.94	10.3	15.1	19.7
ηc(%)	76.2	88.8	91.4	95.6
Xc50(µm)	5.0	2.0	1.5	0.9
α (1)	0.693	0.693	0.693	0.693
m(1)	2.65	1.06	1.06	0.88

FIGURE 53. Fractional collection efficiency curves.

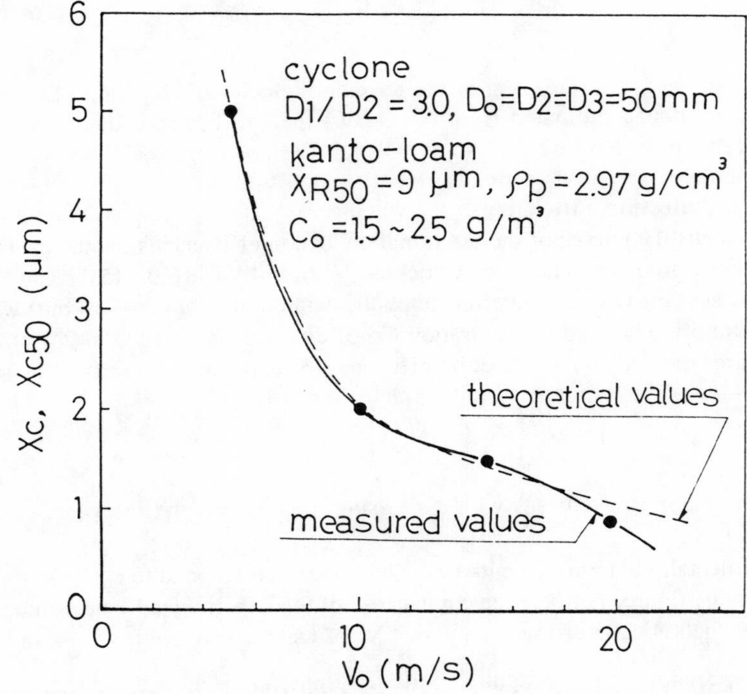

FIGURE 54. Comparison of experimentally determined cut-size with theoretically estimated cut-size of Ogawa's theory.

collected in the body of the cyclone was found deposited on the inside wall of the exhaust pipe.

Abrahamson and Lim devised and tested the "exhaust scrubbing" insert, as shown in Figure 55. The idea behind this insert is to introduce a curtain flow of clean air opposing

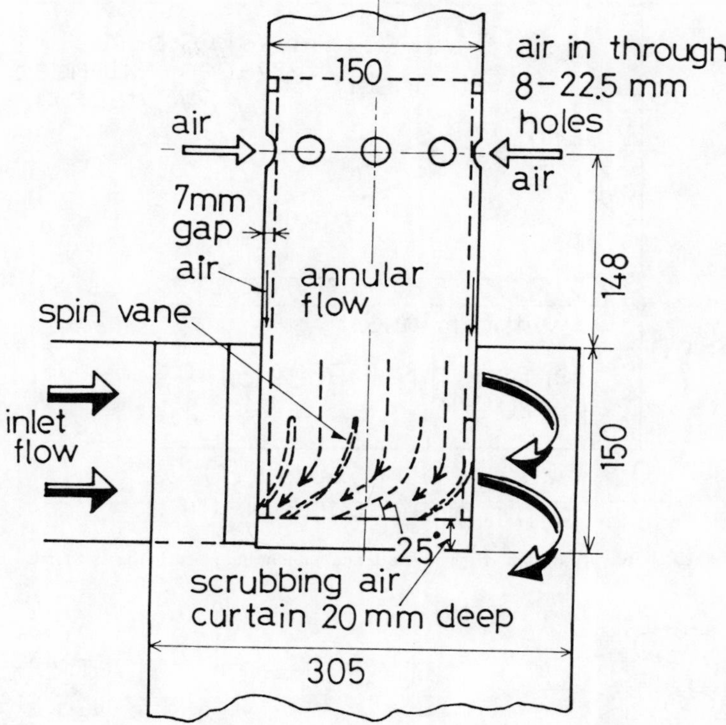

FIGURE 55. Annular air curtain for dust reflux.

the exhaust, flushing deposited dust back into the cyclone. In order to ensure that the entrained dust is carried out into the downward flow where it can be collected, a spin (rotation) velocity is imparted to the curtain flow. This spin velocity was set higher than that calculated for the normal exhaust flow at the lip. Operation of the insert was convenient for the cyclone operating under a vacuum, since no extra fan was required. They used a 305 mm diameter long cone cyclone, similar in geometry to a Stairmand high efficiency type, except that the inlet was narrower; the inlet cross-section was 0.11·D1 × 0.5·D1 rather than 0.2·D1 × 0.5·D1.

Their initial tests were conducted with alumina dust (X_{R50} = 24 μm, ρ_p = 3 kg/m³) with the scrubbing air curtain 20 mm deep and the annular nozzle 7 mm thick. The reverse downward axial velocity was 12 m/s, compared to an average upward exhaust velocity of 6 m/s. The spin (rotational, tangential) velocity of the curtain (12 m/s for tan 25°, where the spin vanes were 25° to the horizontal) was 26.5 m/s. This was 30% higher than the exhaust spin calculated from equations using the exhaust pipe radius as

$$\frac{V\theta}{Vi} = \frac{Ri}{r}\left[1 - \exp\left\{-C\cdot\left(\frac{r}{Ri}\right)^2\right\}\right]$$

where Ri is the radius at the midpoint of the tangential entry area.

With the alumina dust on the feed-dust concentration Co = 4.5 g/m³, the collection efficiency η_c rose from 99.03% with the unmodified cyclone to 99.34% with the insert, i.e., a reduction of about 30% in dust emission. For the same pressure drop there was a reduction in processed air of 8%, and an addition of 45% to the exhaust flow.

Further tests were made with cement dust (X_{R50} = 18 μm, ρ_p = 2.7 g/cm³) and an inlet concentration Co = 20 g/m³. The same insert as before was systematically raised and lowered to determine the optimum height of the air curtain. The experimental results shown

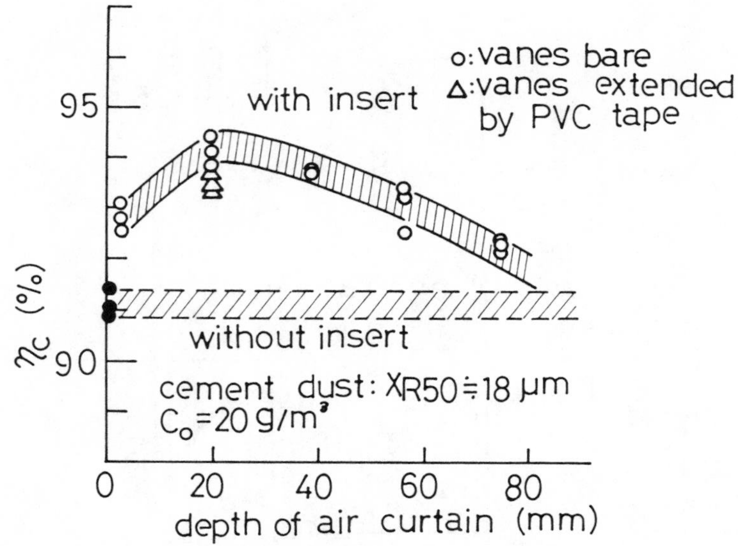

FIGURE 56. Collection efficiency with and without the insert in the exit pipe.

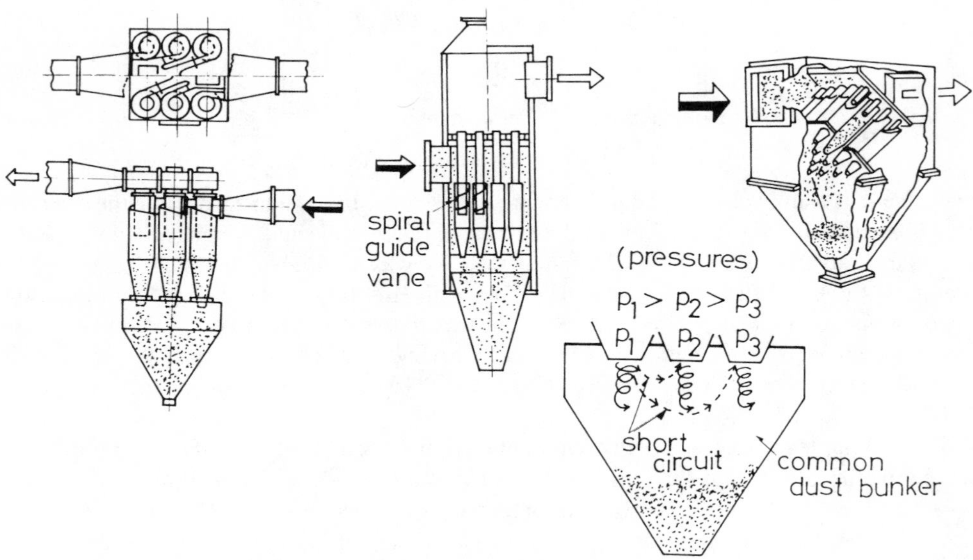

FIGURE 57. Illustrations of three general types of multi-cyclones.

in Figure 56 indicate the optimum chosen position was 20 mm. The results with cement dust are both more accurate and numerous, and indicate a reduction of 35% in dust emission. All velocities were sensibly constant for all runs using the insert. It was noticed that the exhaust pipe was as much as 3 mm out of round, and that gaps of up to 3 mm occurred between the exhaust pipe and the spin vanes. The spin vanes were then extended with plastic tape to cover these gaps, and the efficiency was measured again at the optimum position. The results are also shown in Figure 56 as triangular symbols, and the efficiency has perhaps diminished slightly.

VIII. MULTI-CYCLONES

Figure 57 shows the three general types of multi-cyclones. In general, the collection efficiency of the multi-cyclones is lower than that of the unit cyclone. On the multi-cyclones, the rotational flow of air in the each cyclone is not the same as the intensity of the rotation, but the different intensities of the rotation are due to the different flow rates of air into the each cyclone. Therefore, the static pressure at the exit place for the each cyclone is different, namely, one has lower pressure and another a higher pressure. Then, the additive air current (short circuit) in the common dust bunker is created from the higher pressure to the lower pressure. For this reason, the normal rotational flows of air in the each cyclone are hindered. Consequently, the centrifugal separation decreases due to the short circuit in the common dust bunker.

REFERENCES

1. **Kovvasiuk, A. S.,** *Teploenelgetika,* 6, 30, 1958.
2. **Ter Linden, A. J.,** Investigations into cyclone dust collectors, *Proc. Inst. Mech. Eng.,* 233, 1949.
3. **Silverman, L.,** Performance of inertial cyclones, *Heat. Vent. Eng. J. Air Cond.,* 87, February, 1953.
4. **Bogoslovskij, V. N., Shcheglov, V. P., and Razumov, N. N.,** *Otoplenie i Ventiliatsiia,* Moskva, 1980.
5. **Daniels, T. C.,** Investigation of a vortex air cleaner, *Engineering (London),* 8, 358, 1957.
6. **Heinrich, R. F. and Anderson, J. R.,** Cyclones, in *Chemical Engineering Practice,* Vol. 3, Cremer, H. W. and Davies, T., Eds., Butterworths, London, 1957, chap. 10.
7. British Patent 713,930.
8. **Jotaki, T.,** On the cyclone dust collectors with the guide vanes, *Jpn. Soc. Mech. Eng.,* 20(97), 604, 1954.
9. **Meldau, R.,** Physikalische Eigenschaften von Industriestauben, *Z. Ver. Dtsch. Ing.,* 76(49), 1189, 1932.
10. **Hughes, R. R.,** Use of modern developments in fluid mechanics to aid chemical engineering research, *Ind. Eng. Chem.,* 49(6), 947, 1957.
11. **Jackson, R.,** The flow pattern in cyclones. II., *Br. Coal Util. Res. Assoc.,* 26(7), 221, 1962.
12. **Liu, Q. N., Jia, F., Zhang, D. I., Hao, J. Y., Wang, P. S., and Xu, J. H.,** An experimental study on three-dimensional flows in the cyclone separators, *Acta Mech. Sin.,* 3, 182, 1978.
13. **Ogawa, A.,** Analysis of loci of the fine solid particles and the flow pattern of air near the inlet pipe region in cyclone dust collectors, *Int. Symp. Powder Tech.,* Kyoto, Japan, September 27—October 1, 1981.
14. **Burgers, J. M.,** *Advances in Applied Mechanics,* Vol. 1, Academic Press, New York, 1948, 171.
15. **Reymond, R. F. and Gauvin, W. H.,** Theoretical and experimental studies of confined vortex flow, *Can. J. Chem. Eng.,* 59, 14, 1981.
16. **Casal, J. and Martinez, J. M.,** Calculating pressure drop in cyclones, 2nd Congr. Mediterr. Ingenieria Quimica, Barcelona, November, 1981.
17. **Charlton, J.,** *Textbook of Fluid Dynamics,* Von Nostrand, New York, 1967.
18. **Ogawa, A. and Fujita, Y.,** On the diffusion of fine solid particles and on the intensity of turbulences in a cylindrical cyclone dust collector, 2nd Congr. Mediterr. Ingenieria Quimica, Barcelona, November, 1981.
19. **Kriegel, E.,** *Aufbereit. Tech.,* 9(1), 1, 1968.
20. **Briggs, L. W.,** *Trans. Am. Inst. Chem. Eng.,* 42(3), 511, 1946.
21. **Barth, W.,** *VDI-Berichte,* 6, 29, 1955.
22. **Troiankin, IU. V. and Balueb, E. D.,** *Teploenergetika,* 29, 1969.
23. **Muschelknautz, E.,** Theorie der Fliekraftabscheider mit besonderen Berücksichtigung hoher Temperaturen und Drücke, *VDI-Berichte,* 363, 49, 1980.
24. **Bohnet, M.,** Zyklonabscheider, GVC-Vortrag, *Tech. Gas/Feststoffströmungen,* December, 1981.
25. **Rosin, P., Rammler, E., and Intelmann, W.,** Grundlagen und Grenzen der Zyklonentstaubung, *Z. Ver. Dtsch. Ing.,* 76(16), 433, 1932.
26. **Davies, C. N.,** The separation of airborne dust and particles, *Inst. Mech. Eng. Proc.,* 1B(5), 185, 1952.
27. **Goldshtik, M. A.,** *Inzhenerno-Fizitcheskij Zhurnal,* 5(6), 105, 1962.
28. **Fuchs, N. A.,** *The Mechanics of Aerosols,* Pergamon Press, Oxford, 1964.
29. **Ogawa, A.,** On the theory of the cyclone dust collectors, Proc. Eur. Congr. Transfer Processes in Particle Systems, Nuremberg, March, 1977.
30. **Barth, W. and Leineweber, L.,** Beurteilung und Auslegung von Zyklonabscheidern, *Staub,* 24(2), 41, 1964.
31. **Barth, W.,** Druckverluste und Abscheideleistungen von Zyklonabscheidern, *VDI-Tagungsheft,* 3, 11, 1954.
32. **Abrahamson, J. and Lim, H. K.,** Mechanisms of dust collection in cyclones new developments, Chemeca 81:9th Australiasian Conf. Chem. Eng., Christchurch, New Zealand, August 30—September 4, 1981.
33. **Abrahamson, J.,** *Mechanisms of Dust Collection in Cyclones,* Elsevier, Amsterdam, 1981.

Chapter 2

ROTARY FLOW DUST COLLECTORS

I. INTRODUCTION

Fourteen years ago, the tornado dust collector (Drehströmungsentstauber, or rotary flow dust collector) first appeared as a development in connection with the power plant construction activities of Siemens AG.[1] This type of dust collector could establish the cut-size Xc50 as 0.5 μm or below.[2,3] Therefore we expect a high collection efficiency in comparison with that of ordinary types of cyclones.

The separation mechanism of fine solid particles and the flow pattern of air in a rotary flow dust collector shows very different conditions from those of ordinary types of the cyclones. Figure 1 illustrates the separation mechanism of fine solid particles, the flow systems of the primary dust-laden gas, the secondary circulating air flow, and the total construction of the rotary flow dust collector. The primary dust-laden gas is thrown upward as a swirling jet into the main cylindrical separation chamber from the primary vortex chamber. During entry into the inner pipe of the diameter D2, the fine solid particles are separated by the centrifugal force in the primary vortex flow, and the separated solid particles settle to the dust-bunker following the descending air of the secondary rotational flow.

In order to increase the magnitude of rotation of the primary vortex flow and the effect of the centrifugal separation, secondary pure air is introduced into the main separation chamber of the diameter D1 through two nozzles. This secondary air flows in through the coaxial slit between the inner pipe and the outer pipe, then it re-circulates again to the secondary air nozzles by passing through a blower.

In general, the cut-size Xc50 can be defined as the particle diameter which corresponds to the value of the fractional collection efficiency η_x to be 0.5 (50%), and also the theoretically determined cut-size Xc can be defined as the imaginary particle diameter which can be attained at the edge of the inner pipe located at the height H from the primary vortex chamber, as shown in Figure 1.

In this section, the correlation between a newly derived theoretical cut-size Xc and an experimentally determined cut-size Xc50, and also the effects which are influenced by the flow rates of the primary dust-laden gas Q1 and of the secondary pure air Q2 on the cut-size Xc50, are described in detail.

II. THEORETICAL EQUATION OF THE CUT-SIZE

A. Fundamental Equation for the Cut-Size

The cut-size is defined as the particle diameter that can be reached at the edge (r = R2) of the inner pipe of height H from the radius r of the vortex chamber as shown in Figure 2. Then from a fluid dynamical point of view, we assume that the Reynolds number Rex (Ur·Xp/υ) around a sphere may be regarded as Rex ≦ 4, so that a sphere accepts Stokes drag force for the radial direction.

Here, denoting the velocity vector of the fluid by **V** (Vθ,Vr, Vz) and that of the solid particle by **U** (Uθ, Ur, Uz), the equation of motion for the solid particle in the radial direction can be written

$$\rho_p \frac{\pi \cdot Xp^3}{6} \cdot \left(\frac{dUr}{dt} - \frac{U\theta^2}{r} \right) = -3 \cdot \pi \cdot \eta \cdot Xp \cdot \S \, (Ur - Vr) \qquad (1)$$

where suffixes θ, r, and z denote the tangential, radial, and axial components, respectively.

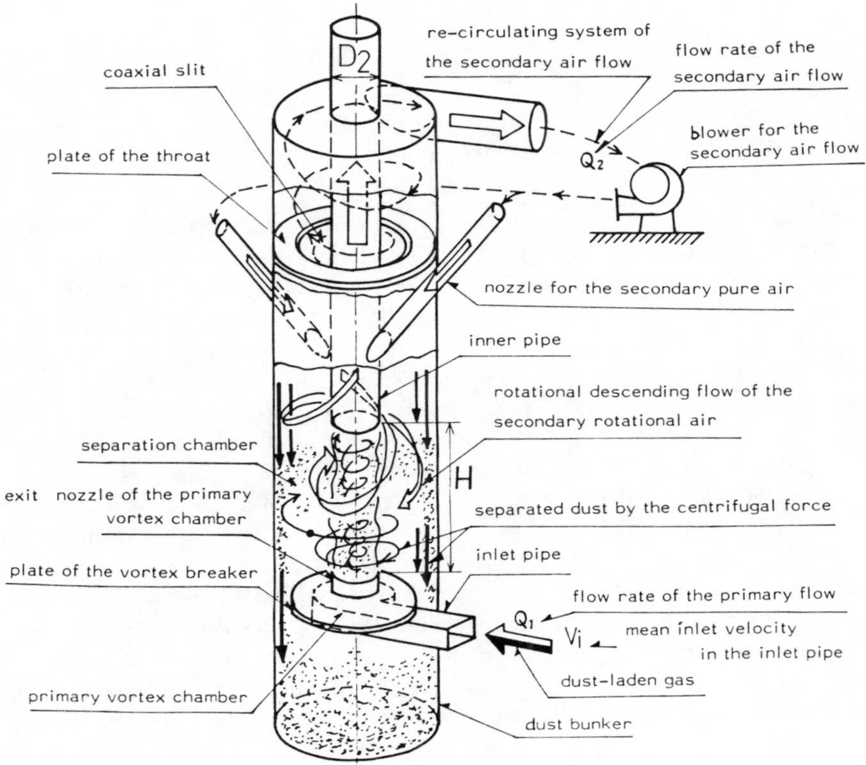

FIGURE 1. Illustration for flow systems and for the separation mechanism of a rotary flow dust collector.

In order to simplify the above equation, assume that the solid particles are in a quasi-steady motion in the radial direction, and the term dUr/dt is nearly zero. Therefore, Equation 1 becomes

$$\rho_p \cdot \frac{\pi \cdot Xp^3}{6} \cdot \frac{U\theta^2}{r} = 3 \cdot \pi \cdot \eta \cdot Xp \cdot \S\,(Ur - Vr) \tag{2}$$

Then the rotation of the primary vortex flow mainly depends on the ratios of the flow rates Q2/Q1, but considering a flow model of the rotational cylindrical shell which flows upward, the radial velocity of flow in the primary vortex flow is assumed zero as Vr ≒ 0. Then Equation 2 becomes

$$p_p \cdot \frac{\pi \cdot Xp^3}{6} \cdot \frac{U\theta^2}{r} \doteqdot 3 \cdot \pi \cdot \eta \cdot Xp \cdot \S \cdot Ur \tag{3}$$

In this equation, the radial velocity Ur of the solid particle can be transformed as

$$Ur = \frac{dr}{dt} = \frac{dr}{dz} \cdot \frac{dz}{dt} = Uz \cdot \frac{dr}{dz} \tag{4}$$

so Equation 3 becomes

$$Uz \cdot \frac{dr}{dz} = \frac{\rho_p \cdot Xp^2}{18 \cdot \eta \cdot \S} \cdot \frac{U\theta^2}{r} \tag{5}$$

Further, it is assumed that the tangential and axial velocities of solid particles are equal to those of the fluid velocities, and also that the axial velocity Vz of the primary vortex flow

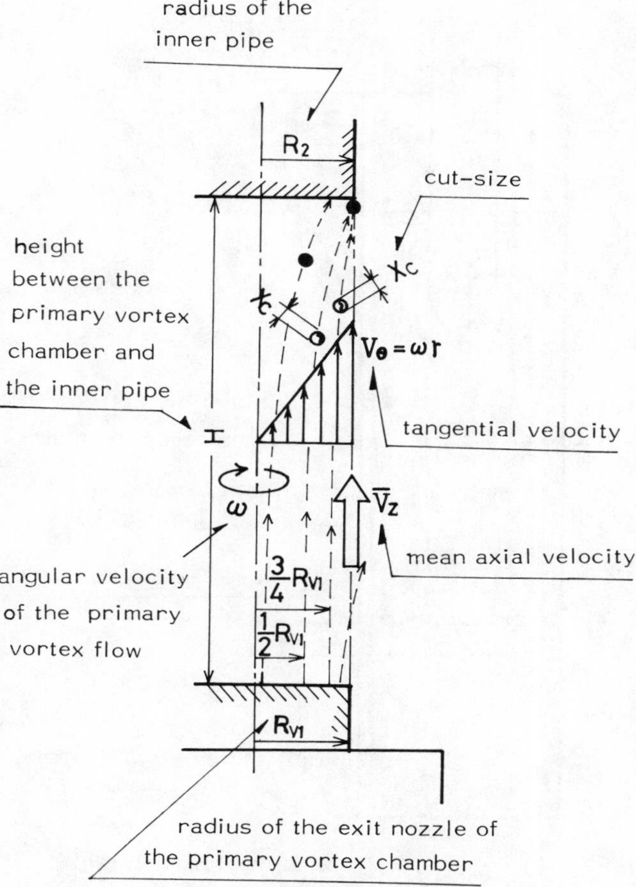

FIGURE 2. Illustration for defining the cut-size.

goes up nearly with a constant velocity Vz from the vortex chamber to the edge of the inner pipe. From the experimental results of the tangential velocities in the primary vortex flow, the tangential velocity Vθ may be written approximately as follows:

$$V\theta = \omega \cdot r \tag{6}$$

Substituting Equation 6 into Equation 5, the equation of the cut-size Xc which satisfies the boundary conditions (starting position, z = 0, r = r; final position, z = H, r = R2) can be obtained as follows:

$$\ln \frac{R2}{r} = \frac{\rho_p \cdot Xc^2 \cdot \omega^2 \cdot H}{18 \cdot \eta \cdot \S \cdot \bar{V}z} \tag{7}$$

Therefore Equation 7 can be transformed as

$$Xc(r) = \sqrt{\frac{18 \cdot \eta \cdot \S \cdot \bar{V}z}{\rho_p \cdot \omega^2 \cdot H} \cdot \ln \frac{R2}{r}} \tag{8}$$

where Xc(r) denotes the cut-size corresponding to the starting radial position at r = r and z = 0, where § is the shape factor of the solid particle, η is the viscosity of gas, ρ_p is the density of the solid particles, and ω is the angular velocity of the primary vortex flow.

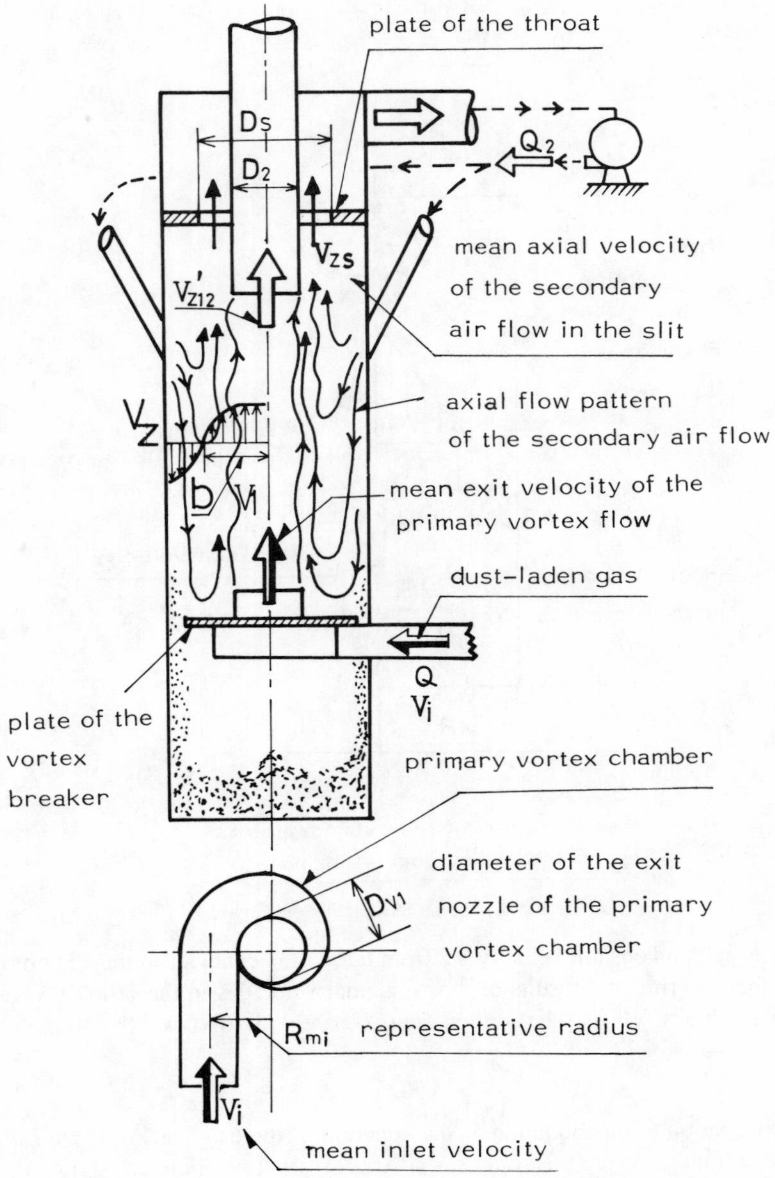

FIGURE 3. Illustration of the primary vortex chamber and of the main axial velocities.

B. Relationship Between the Tangential Velocity of the Primary Vortex Flow and the Construction of the Primary Vortex Chamber

As shown in Figure 3, denoting by Rmi the representative radius of the vortex chamber inlet pipe and by Vi the mean inlet velocity in the inlet pipe, the relationship between the radius r and the tangential velocity Vθ near the exit radius Rv1 of the vortex chamber may be described in the region where the theorem of the conservation of the angular momentum is applicable as follows:

$$K1 \cdot V\theta \cdot r = Rmi \cdot Vi \tag{9}$$

where K1 is a constant which can be determined by the ratios of the primary flow rate Q1 and the secondary flow rate Q2. Then the relationship between K1, Q1 and Q2 can be written experimentally as

$$K1 = \Phi(Q1/Q2) \tag{10}$$

Therefore, the tangential velocity Vθv of the primary vortex flow at the exit nozzle of the radius Rv1 can be written

$$V\theta v = \frac{Rmi}{K1} \cdot \frac{Vi}{Rv1} = \frac{Rmi}{Rvi} \cdot \frac{Q2}{Q1} \cdot \frac{Vi}{\Phi} \tag{11}$$

so the equation for the angular velocity ω of the primary vortex flow can be written

$$\omega = \frac{1}{\Phi} \cdot \frac{Rmi}{Rv1} \cdot \frac{Q2}{Q1} \cdot \frac{Vi}{Rv1} \tag{12}$$

C. Representative Axial Velocity of the Primary Vortex Flow

As shown in Figure 3, the mean axial velocity V1 at the exit nozzle of the primary vortex chamber can be written as $V1 = 4 \cdot Q1/\pi \cdot Dv_1^2$ and the mean axial velocity $V\acute{z}12$ of the primary flow entering the inlet pipe of the diameter D2 can be written as $V\acute{z}12 = 4 \cdot Q1/\pi \cdot D_2^2$. Then denoting the flow rate of the secondary air by Q2, the mean axial velocity Vzs of the secondary air flowing through the slit between the diameters Ds and D2 can be written

$$Vzs = \frac{4 \cdot Q2}{\pi \cdot (Ds^2 - D2^2)}$$

Consequently, the mean axial velocity of the primary vortex flow near the inner pipe may be assumed as

$$\overline{V}z12 = \frac{(V\acute{z}12 + Vzs)}{2}$$

This equation can be rewritten as

$$\overline{V}z12 = \frac{V\acute{z}12 + Vzs}{2} = \frac{2 \cdot Q1}{\pi \cdot D2^2} \cdot \left\{1 + \frac{(Q2/Q1)}{(Ds/D2)^2 - 1}\right\} \tag{13}$$

Therefore, the mean axial velocity $\overline{V}z$ along the effective height H can be written

$$\overline{V}z12 = \frac{2 \cdot Q1}{\pi \cdot Dv1^2}\left[1 + \frac{1}{2} \cdot \left(\frac{Dv1}{D2}\right)^2 \cdot \left\{1 + \frac{(Q2/Q1)}{(Ds/D2)^2 - 1}\right\}\right] \tag{14}$$

D. Theoretical Formulas for the Cut-Sizes

Substituting Equations 12 and 14 into Equation 8, and arranging Equation 8, the following equation can be obtained as

$$Xc(r) = \Phi \cdot \frac{Rv1}{Rmi} \cdot \frac{Q1}{Q2} \cdot \frac{3}{Vi} \cdot \sqrt{\frac{\S \cdot \eta}{\rho_p} \cdot \frac{Q1}{\pi \cdot H} \cdot \left[1 + \frac{1}{2} \cdot \left(\frac{Dv}{D2}\right)^2 \cdot \left\{1 + \frac{(Q2/Q1)}{(Ds/D2)^2 - 1}\right\}\right] \cdot \ln\frac{Rv1}{r}} \tag{15}$$

Defining the starting radius positions at the exit nozzle of the primary vortex chamber as r = Rv1/2 and r = 3 Rv1/4, the following equations of the cut-size can be obtained by Ogawa[4] as

$$Xc\left(\frac{1}{2} Rv1\right) = \Phi \cdot \frac{Rv1}{Rmi} \cdot \frac{Q1}{Q2} \cdot \frac{3}{Vi} \cdot \sqrt{\frac{\S \eta Q1}{\rho_p \pi H} \cdot \left[1 + \frac{1}{2} \cdot \left(\frac{Dv1}{D2}\right)^2 \cdot \left\{1 + \frac{(Q2/Q1)}{(Ds/D2)^2 - 1}\right\}\right] \cdot \ln 2} \tag{16}$$

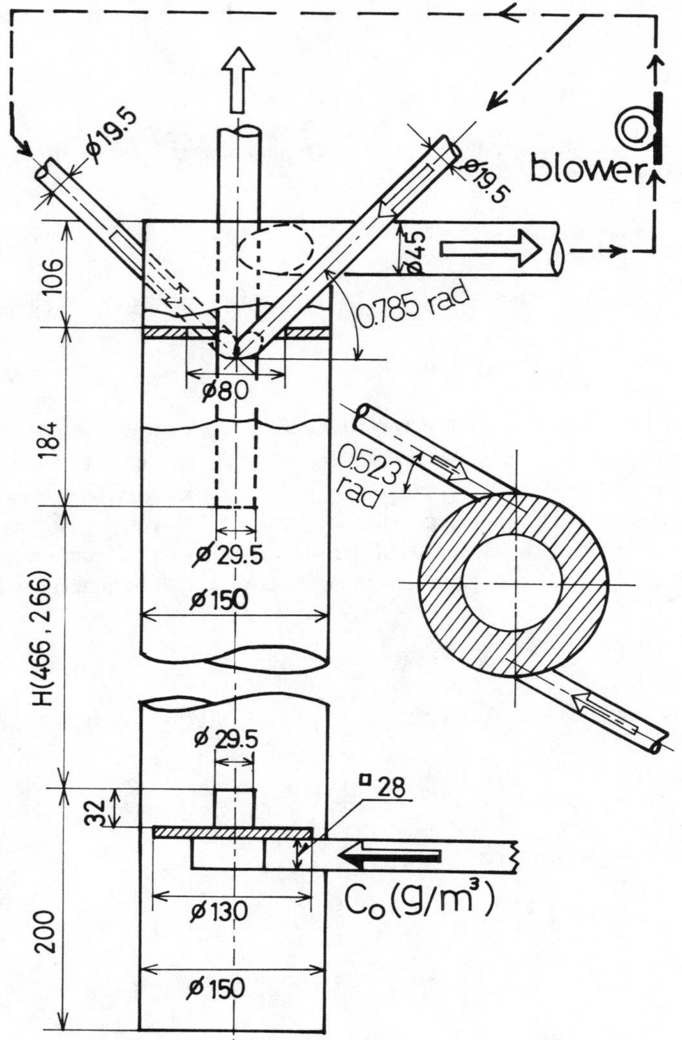

FIGURE 4. Main sizes of a rotary flow dust collector.

$$Xc\left(\frac{3}{4}Rv1\right) = \Phi \cdot \frac{Rv1}{Rmi} \cdot \frac{Q1}{Q2} \cdot \frac{3}{Vi} \cdot \sqrt{\frac{\S \eta Q1}{\rho_p \pi H} \cdot \left[1 + \frac{1}{2} \cdot \left(\frac{Dv1}{D2}\right)^2 \cdot \left\{1 + \frac{(Q2/Q1)}{(Ds/D2)^2 - 1}\right\}\right] \cdot \ln\frac{4}{3}} \tag{17}$$

III. NUMERICAL EXAMPLE OF CUT-SIZE

A detailed construction of the rotary flow dust collector is shown in Figure 4, and the values of the cut-sizes can be calculated by Equations 16 and 17. The necessary numerical values are Rmi = 28.8 mm, Dv1 = 29.5 mm, D2 = 29.5 mm, Ds = 80 mm, H = 466 mm, the density of fly-ash ρ_p = 2.14 g/cm^3 and the viscosity of air η = 1.814 $\times$ 10^{-5} Pa·s, and the value of Φ could be determined as Φ = 1.6 by the experimental results of the tangential velocities. Also a shape factor § of the solid particle is supposed as § = 1. The calculated results of the cut-sizes are shown in Figure 5 for the cases of Q2/Q1 = 0.2, 0.5, 1.0, 1.5, and 2.0, and also in this figure the cut-sizes Xc (r = Rv1/2) and Xc (r = 3·Rv1/4) are shown as the solid lines and the dotted lines, respectively.

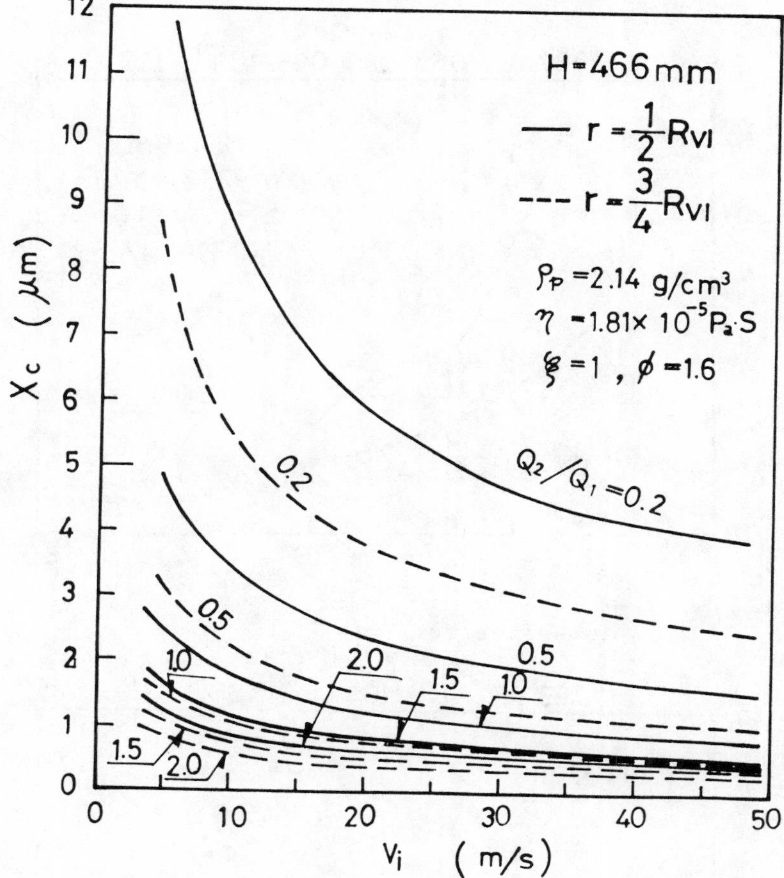

FIGURE 5. Calculated results of the cut-size for H = 466 mm.

IV. COMPARISON OF THE EXPERIMENTAL RESULTS OF CUT-SIZES Xc50 WITH THOSE OF Xc

Here, the relationships between the calculated cut-size Xc and the experimentally determined cut-size Xc50, which can be estimated by the fractional collection efficiency, are discussed. Figure 6 (a) and (b) show the cut-sizes for H = 466 mm and 266 mm, respectively. The starting radial positions in the exit nozzle of the primary vortex chamber are r = Rv1/2 and r = 3·Rv1/4, respectively. The comparisons of the calculated cut-size Xc with the experimentally determined cut-size Xc50 for fly-ash as a test dust are shown in these figures. From these results, it is sufficiently possible to estimate the cut-size Xc by Equations 16 and 17 in comparison with the experimentally determined cut-size Xc50 in the domain Q2/Q1 = 0.9 to 1.6.

Because the actual flow pattern in this collector coincides with the assumed flow pattern in the domain Q2/Q1 = 1.0 to 1.5 in connection with the tangential velocities Vθ and the axial velocities Vz which are concerned with the flow rates Q1 and Q2. It is made especially clear that the separation mechanism for fine solid particles is largely hindered; also the pressure drop is increased by the generation of eddies due to the velocity differences at the mixing fluid layers where the flow of the primary vortex layer contacts with the secondary air flow.

Consequently, for the derivation of these equations, the ratios Q2/Q1 of the flow rates Q2 and Q1 are the most important factors concerning the tangential and axial velocities.

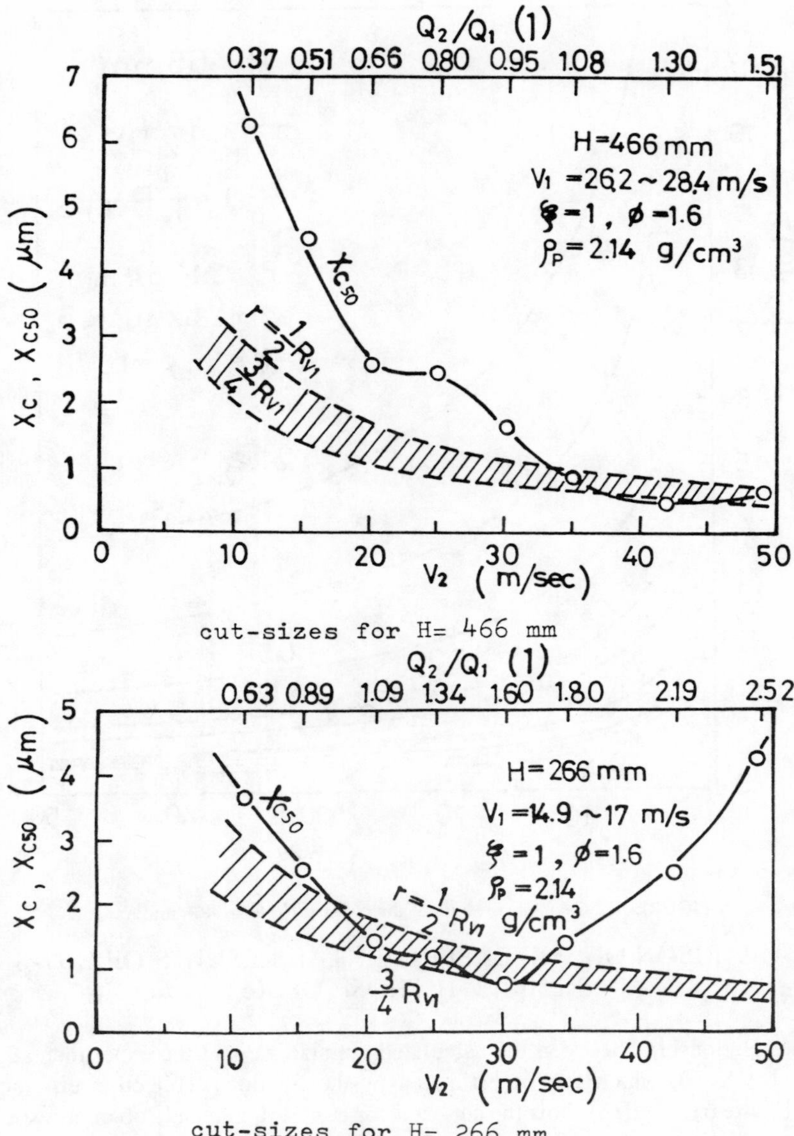

FIGURE 6. Relationships between Xc and Xc50 for a rotary flow dust collector.

Then, one interesting fact is that the cut-sizes Xc are independent of the outer diameter D1 by Equations 16 and 17. Also the coincidence between the calculated values Xc and the experimentally determined cut-sizes Xc50 is satisfied in the range of Q2/Q1 = 0.9 to 1.6. Still more from a point of view of the pressure drop, this range of the ratios of the flow rates is very useful.

V. EXPERIMENTAL APPARATUS AND EXPERIMENTAL METHOD

As shown in Figure 7, the rotary flow dust collector is made of a transparent synthetic resin. The main dimensions are the outer cylinder diameter: D1 = 150 mm, the diameter of the inlet pipe where the dust-laden air flows into the main body of the vortex chamber: 29.5 mm, and the diameter of the outlet pipe from the main body of vortex chamber: 29.5

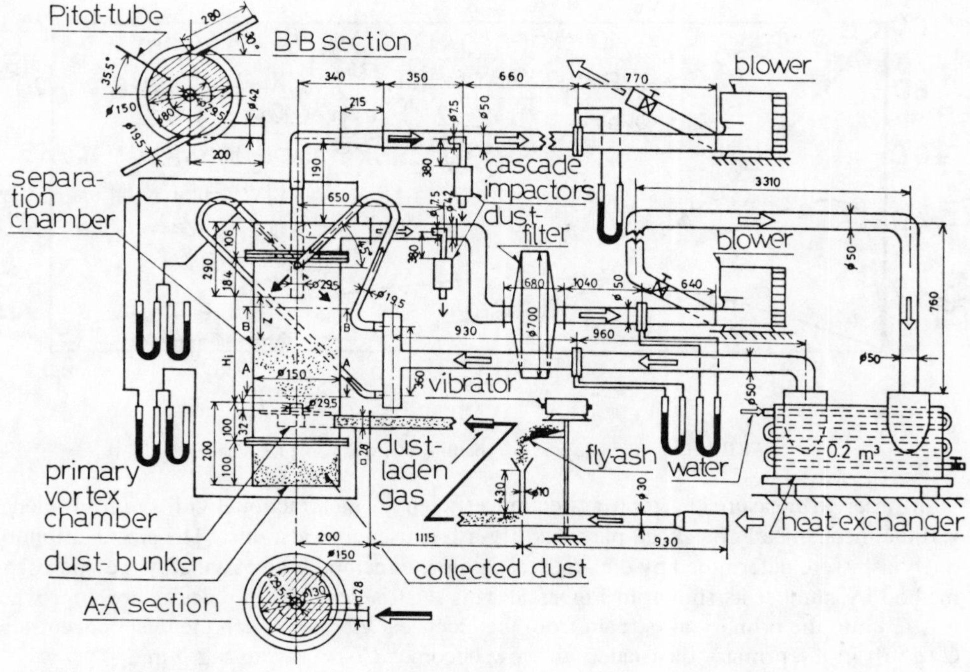

FIGURE 7. Experimental apparatus and the total flow system.

mm. The tangential inlet pipe of the secondary air flow has the diameter 19.5 mm. This pipe is installed with a slope of 45° to the collector center line and is inclined at 30° in the tangential direction.

The secondary air flow streams into the main body of the vortex chamber from two inlet pipes which are installed symmetric with respect to the axis. The diameter of the exit slit above the main body of the vortex chamber for the secondary air flow is 80 mm. The exit pipe of the primary air flow is installed passing through the center of this exit slit. The diameter of the exit pipe for the secondary air flow from the upper part of the collector is 42 mm. The diameter of the interruptive disc plate above the dust-bunker is 130 mm.

The height of the outlet pipe from the primary vortex chamber is 32 mm. The secondary air flow re-circulating in the pipe is controlled by a heat-exchanger of volume 0.2 m^3, through which the heated air flowing through the friction blower is cooled to the normal room temperature.

In order to compare the effects of the effective height H on the performance of this collector, two kinds of H are employed, H = 266 mm and H = 466 mm.

The velocity distributions of the rotational flow in the case of H = 466 mm are measured by a cylindrical Pitot-tube of diameter 3 mm with a measuring hole of a diameter of 0.3 mm which is inserted perpendicular to the center axis of the collector and also variable to seven Z places for the measurement. The measurable range of velocity distributions of the rotational air flow in the collector is restricted to V1 = 10.1 to 33.1 m/s and V2 = 0 to 49.1 m/s. Then from the results of the axial velocities, equi-flow rate lines are calculated. Each pressure drop for the primary air flow and for the secondary air flow of the collector is measured in the range of V1 = 6.1 to 42.3 m/s and V2 = 0 to 49.1 m/s. The total pressure drop is calculated by considering the flow rate ratio Q2/Q1 and each dynamic pressure.

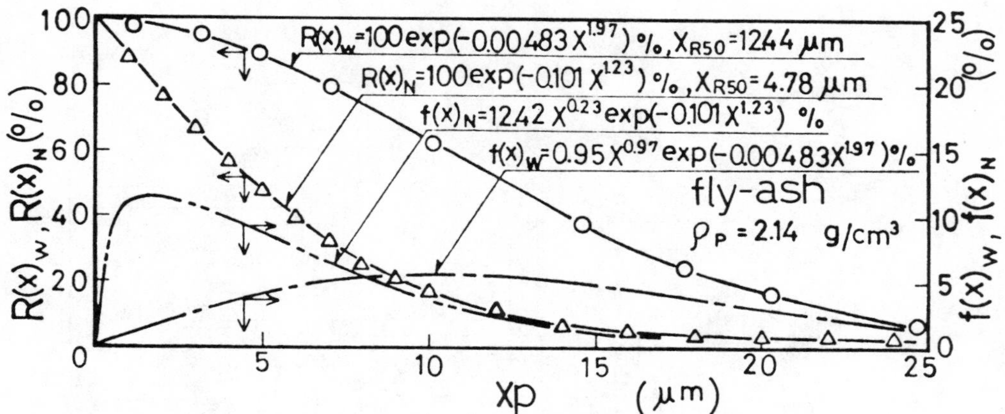

FIGURE 8. Particle size distributions of fly-ash as a test dust.

In order to measure the total collection efficiency, the fractional collection efficiency, and the mechanically balanced particles, fly-ash is used as a test dust. The size distributions of fly-ash were determined by the Andreasen pipette method by mass and by the microscope method by number as shown in Figure 8. This dust is dried up in a desiccator after which it is fed into the primary air stream from the electrical vibrator. Then the dust concentration C (g/m^3) in the primary dust-laden air flow becomes Co = 4.6 to 5.2 g/m^3.

The total collection efficiency is determined by the ratio of total collected dust mass and the total fed dust mass. The fractional collection efficiency which is derived from the method of Schmidt as described below, is determined from the size distribution measured with a five-stage cascade-impactor into which escaping dust is sucked by an isokinetic sampling method of a moving gas stream with a suction probe of an inner diameter of 7.5 mm and an outer diameter of 9.0 mm located at each center of the two exit pipes of the collector. Then the mechanically balanced particles on the arbitrary radius r are determined from the size distributions of dust which are stuck on the adhesive tape on a piano steel wire of diameter 2 mm in the case of the feed concentration Co = 0.4 to 1.2 g/m^3. The range of the inlet velocity of these experiments for the total collection efficiency, the fractional collection efficiency and the mechanically balanced particles are V1 = 10.1 to 33.1 m/s and V2 = 11.1 to 49.1 m/s.

VI. EXPERIMENTAL RESULTS

A. Velocity Distributions and Equi-Flow Rate Lines

The tangential velocity contributing to the dust separation increases with an increasing velocity V2 and then the velocity index n describing the distribution of the tangential velocities for the quasi-free vortex flow was less than about 0.5. This index n became lower in comparison with the general rotational air flow. In addition, index n became lower than about 0.2 in the case of values larger than Q2/Q1 = 1. The decreasing tendency of this value n increased with an increasing value of V2. The effect of the decreasing tendency of index n was strongly influenced at the upper portion of $Z \leqq 200$ mm.

In Figure 9A the distribution of the tangential velocities corresponding to the change of the value of V2 in the case of V1 = 25.8 to 28.4 m/s is shown at the position Z = 200 mm. In these results, in the range of Q2/Q1 = 0.95 to 1.51 the value of n becomes n = 0.07 to 0.15. In the case of Q2/Q1 = 0.37 to 0.66, the value of n becomes n = 0.23 to 0.41. The distribution of the axial velocities corresponding to the values above mentioned range is shown in Figure 9B.

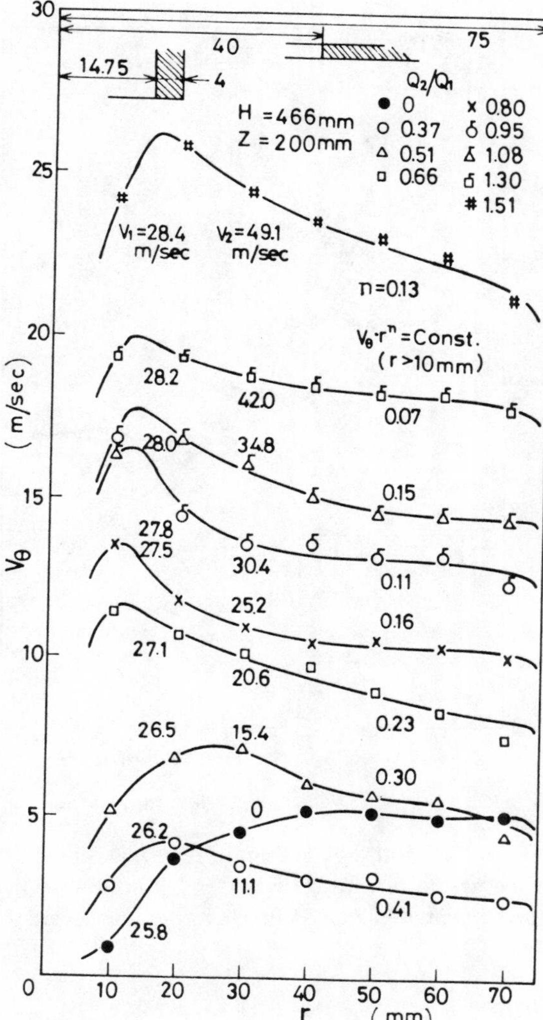

FIGURE 9. (A) Distributions of the tangential velocities.

The radial positions (r = b) where downward flow turned to upward flow (the axial velocity Vz = 0) were in the range of 50 to 60 mm, and these positions became about 50 mm with an increasing velocity of V2. These radial positions became b = 30 to 40 mm. These were almost in the central part from the center axis of the vortex chamber to the inner wall surface in the case of V1 = 14.9 to 17.0 m/s at Z = 340 mm and also b = 40 to 50 mm in the case of V1 = 26.2 to 28.4 m/s. Then at the axial position Z = 8 mm, b became 60 to 70 mm. Below the velocity of V2 = 15 m/s and above V1 = 30 m/s, the radial position of b increased to b = 70 mm and showed a decreasing downward flow.

From these distributions of the axial direction, the dimensionless equi-flow rate lines ψ were calculated as

$$\Psi = \frac{1}{Q1}\int_0^r 2\cdot\pi\cdot r\cdot dr\cdot Vz \tag{18}$$

In the case of Q2/Q1 = 0.80 of Figure 10(A), the equi-line of ψ = 1.0 returned to above the middle of the main body of vortex chamber, but in the case of Q2/Q1 = 1.30 of Figure 10(B), the equi-line of ψ = 1.0 extended to the bottom of the vortex chamber. So in the

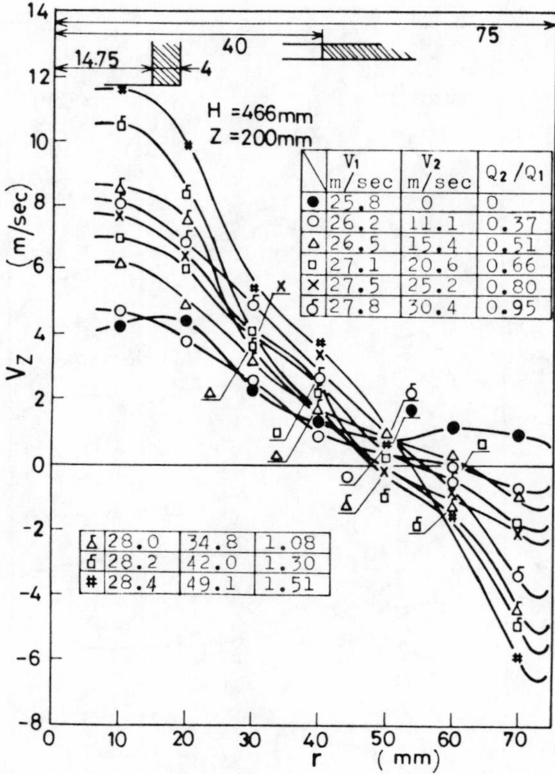

FIGURE 9. (B) Distributions of the axial velocities.

case of Q2/Q1 = 1.30, the inward drag acting on the separating particles which reach the wall surface with an increasing downward flow tends to decrease in comparison with the centrifugal force by the tangential velocity Vθ. Therefore, the separation efficiency becomes higher.[5]

B. Total Pressure Drop

The pressure drop Δp_1 of the primary air flow is defined as

$$\Delta p_1 = \Delta p_{s1} + \rho \cdot \frac{V'1^2}{2} - \rho \cdot \frac{V1e^2}{2} \tag{19}$$

and also the pressure drop Δp_2 of the secondary air flow is defined as

$$\Delta p_2 = \Delta p_{s2} + \rho \cdot \frac{V2^2}{2} - \rho \cdot \frac{V2e^2}{2} \tag{20}$$

Therefore, considering the re-circulation of the secondary air flow, the total pressure drop Δp_c is estimated as[6]

$$\Delta p_c = \Delta p_1 + \left(\frac{Q2}{Q1}\right) \cdot \Delta p_2 \tag{21}$$

The value of Δp for the pure air flow calculated in the case of H = 466 mm is shown in Figure 11. The corresponding minimum values of Δp_c for the various values of V2 were about Q2/Q1 ≒ 1.2. These values of Δp_c were independent of the feed dust concentration Co = 4.6 to 5.2 g/m^3 of fly-ash.

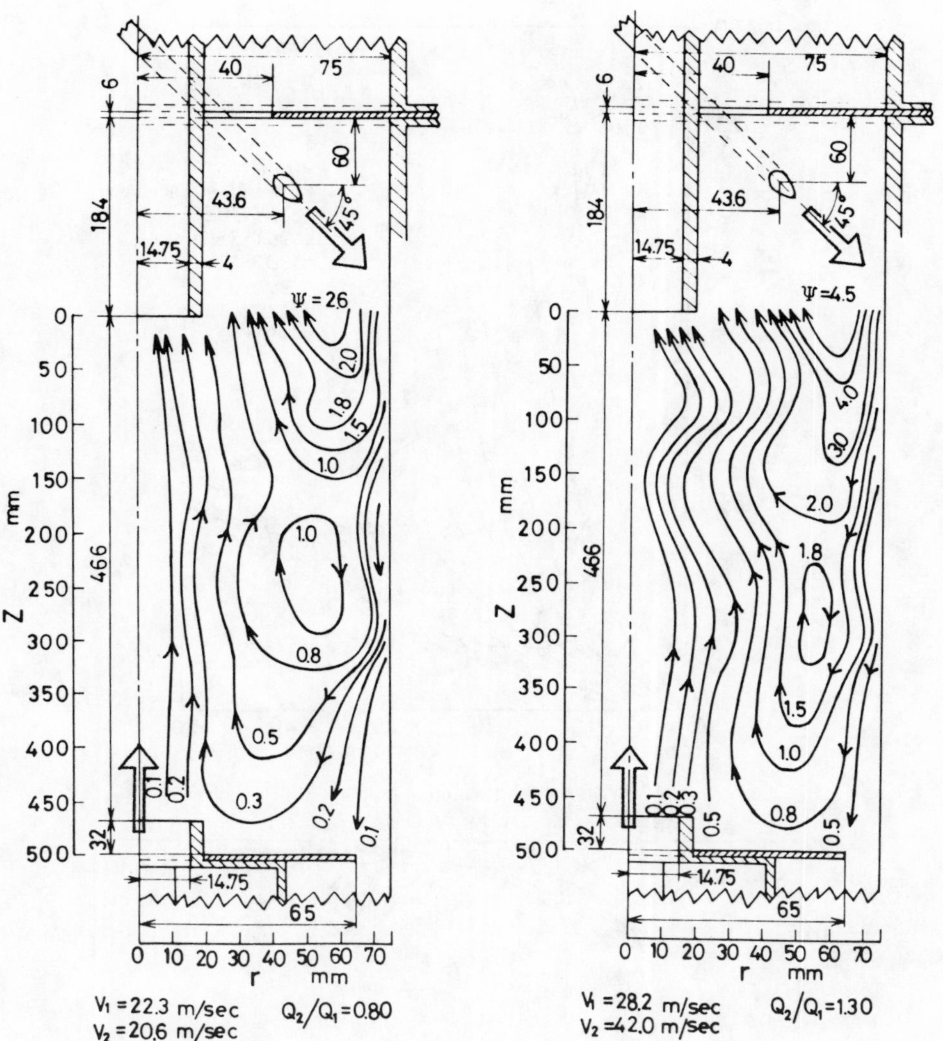

FIGURE 10. (Left) Dimensionless equi-flow rate lines for Q2/Q1 = 0.80. (Right) Dimensionless equi-flow rate lines for Q2/Q1 = 1.30.

C. Total Collection Efficiency

The experimental values of η_c for V2 are shown in Figure 12 and Figure 13 as a parameter of value V1. These values of η_c were very stable and also could reappear within the range of $\pm 3\%$. The value of V1 increased with an increasing value of V2 from the minimum value to the maximum value within each range of V1 which is written in these figures.

In the case of H = 466 mm the value of η_c was more than 95% for the velocity higher than V2 = 20 m/s. Then the driving conditions for the collection efficiency corresponded to the velocity V2 = 42 m/s for the primary inlet velocities V1 = 28.2 and 32.7 m/s, therefore the values of Q2/Q1 were 1.30 in the former case and 1.12 in the latter case. The critical point of a sharply decreasing collection efficiency corresponded to the value Q2/Q1 $\fallingdotseq$ 1.0.

In the case of H = 266 mm which increases the value of V1 from 26 m/s to 33 m/s, the value of η_c decreased about 5 to 10%. The physical meaning of this result is based upon the short residual time of the fine solid particles, and in this case even if the same order of

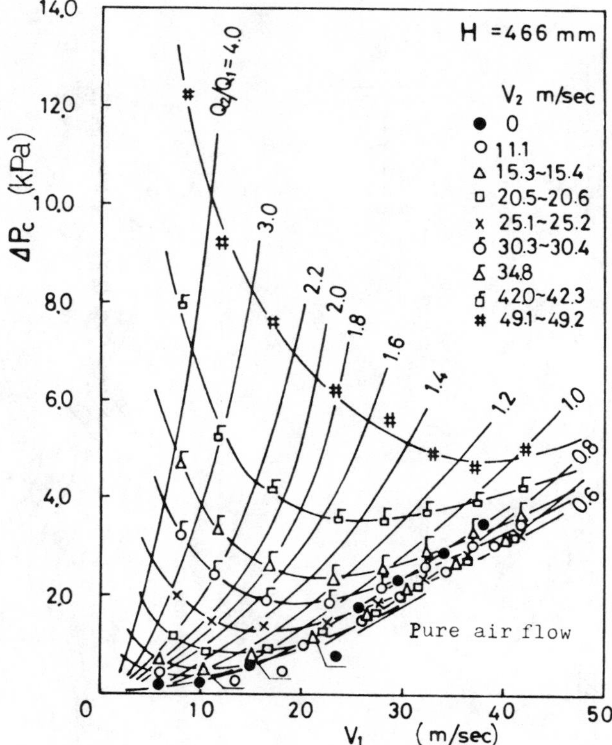

FIGURE 11. Total pressure drop.

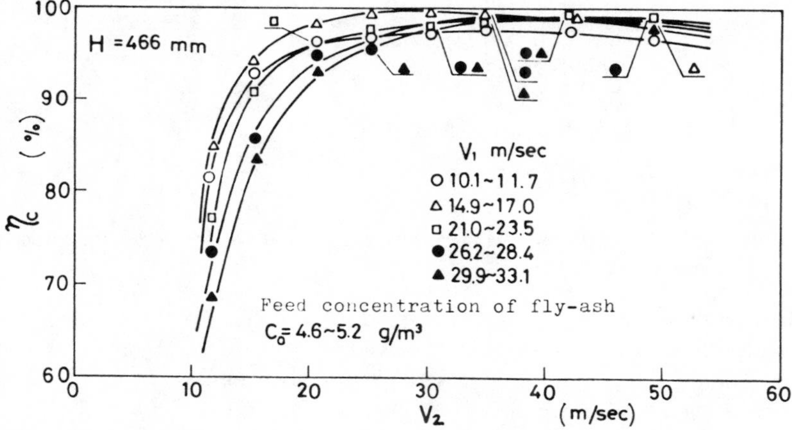

FIGURE 12. Total collection efficiency for the secondary inlet velocity V2 in the case of H = 466 mm.

magnitude of the centrifugal force in the case of H = 466 mm is given, the inward drag acting on the fine solid particles by the returned flow of the secondary air flow acts strongly.

Then the lines of the equi-total collection efficiency η_c are shown in Figure 14 (H = 466 mm) and Figure 15 (H = 266 mm) on the axis V1 and V2 as a parameter Δp_c. In these figures, the lines of the equi-total pressure drop as a parameter is shown. The domain of the total collection efficiency above η_c = 99% for H = 466 mm range in V1 = 13 to 35 m/s and in V2 = 25 to 53 m/s; in this case the value of Q2/Q1 is above about 0.95 and the minimum value of Δp_c is about 1.30 kPa. In comparison with the case of H = 466 mm

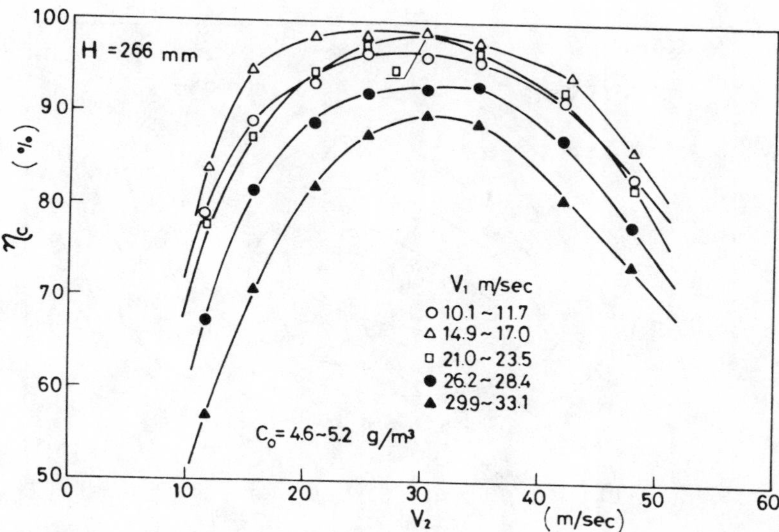

FIGURE 13. Total collection efficiency for the secondary inlet velocity V2 in the case of H = 266 mm.

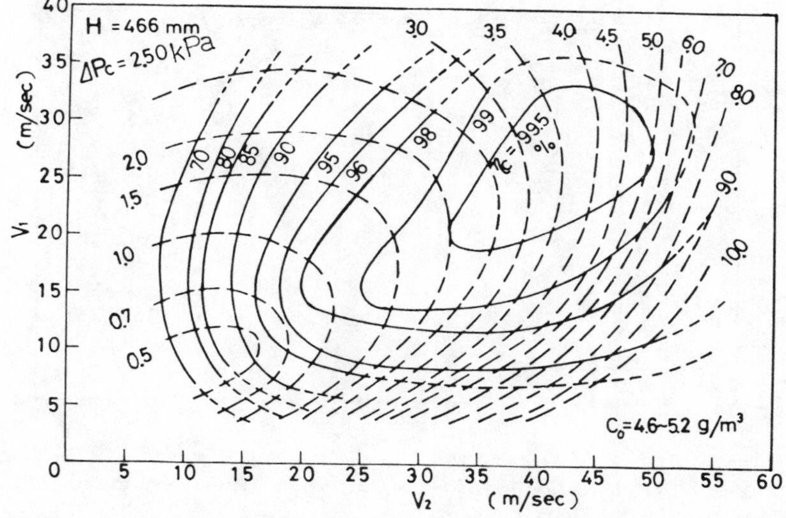

FIGURE 14. Equi-total collection efficiency lines and equi-total pressure drop lines in the case of H = 466 mm.

for the case of H = 266 mm, the domain exhibiting a high collection efficiency was narrower but kept about η_c = 98% even in the case of Δp_c = 0.80 to 1.50 kPa.

In addition to this, the values of η_c for the total consumed energy E = Q1·Δp_c (= Q1·Δp_l + Q2·Δp_2) *N*–m/s of the pure air flow are shown as a parameter Q2/Q1 in Figures 16 and 17.

In the case of H = 466 mm which increases the value of E, η_c developed until E ≒ 78 *N*–m/s. For the case of Q2/Q1 ⩾ 1, η_c = 98% was shown at about E ≒ 9.8 *N*–m/s. When the value of E was increased for H = 266 mm, the value of η_c did not increase proportionally to the value of E, and also the maximum value of η_c at E = 9.8 to 30.0 *N*–m/s was shown. Then with a further increase of the value of E, the value of η_c decreased gradually.

D. Fractional Collection Efficiency

The value of the fractional collection efficiency η_x was calculated by the method of Schmidt

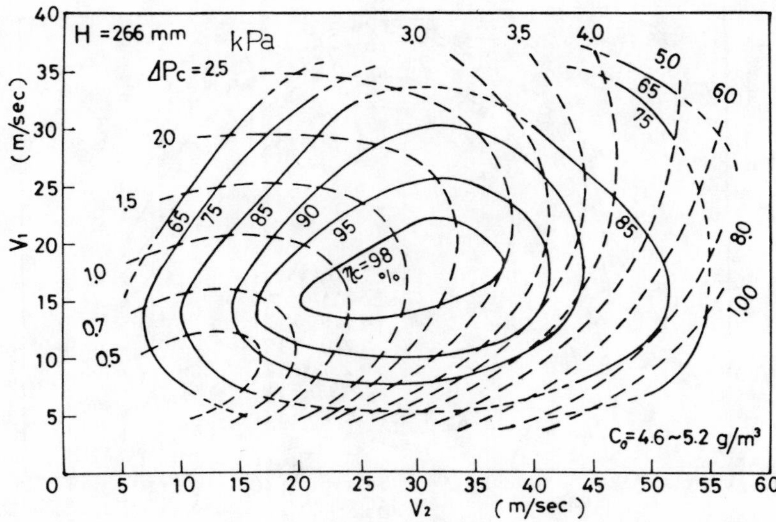

FIGURE 15. Equi-total collection efficiency lines and equi-total pressure drop lines in the case of H = 266 mm.

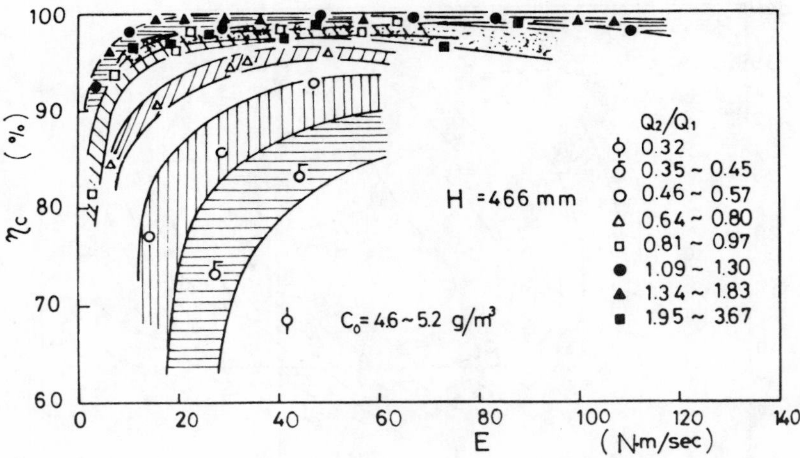

FIGURE 16. Total collection efficiency for the consumption of energy in the case of H = 466 mm.

from which the relationship between the distribution f(Xp)*me* of the escaping particles from the collector and the distribution f(Xp)*m* of the fly-ash feed dust was determined by experiment. This reciprocal relation among η_x, η_c, f(Xp)*me* and f(Xp)*m* is written

$$\eta_x = 1 - (1 - \eta_c)\cdot\left\{\frac{f(Xp)me}{f(Xp)m}\right\}_x \quad (22)$$

Figure 18 (A) and (B) and also Figure 19 (A) and (B) show the experimental results of the fractional collection efficiency η_x for the case of V1 = 14.9 to 17.0 m/s and of V1 = 26.2 to 28.4 m/s, both of which are dependent on the change of V2. After arrival at η_c = 99%, the value of η_x reached about 100% for the particle size Xp = 3 to 5 μm and also in the case of the maximum value of η_c = 99.7%, η_x became about 100% for about Xp ≒ 2.5 μm. For η_c = 95%, η_x became zero in the range of roughly Xp = 1 to 1.5 μm. In the case of particles finer than Xp = 1 to 1.5 μm, {f(Xp)*me*/f(Xp)*m*}x ≧ 20 was given from

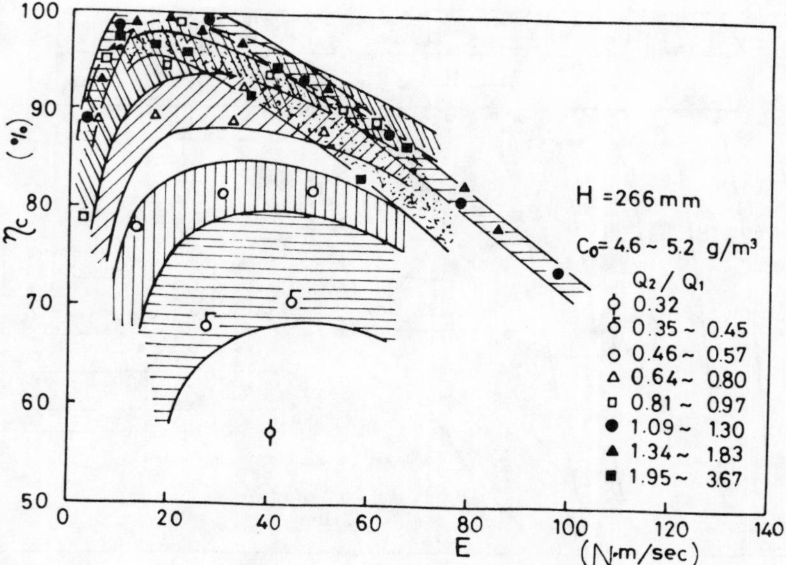

FIGURE 17. Total collection efficiency for the consumption of energy in the case of H = 266 mm.

the results of the experiments for $1 - \eta_c = 0.05$ in Equation 22 and then the values of η_x came up to the maximum of zero. Therefore, these calculated results indicate that these fine particles cannot be collected. By observing under the microscope the fly-ash which was collected in the dust bunker, even the finer particles of lower size than Xp = 1 to 1.5 μm were confirmed sufficiently.

On the other hand, in the case of $\eta_c = 67.3\%$ for V1 = 26.2 m/s, V2 = 11.1 m/s, Q2/Q1 = 0.37, and H = 266 mm, the value of η_x became zero for about Xp ≒ 4.2 μm and reached $\eta_x = 96\%$ for Xp = 15 μm.

E. Cut-Size and Separation Index

The empirical equation of the fractional collection efficiency η_x was written

$$\eta_x = 1 - \exp\left\{(-\ln \cdot 2)\left(\frac{Xp}{Xc50}\right)^m\right\} \qquad (23)$$

and then the collection efficiency was estimated by the factors of the cut-size Xc50 and the index m of the separation (which indicates the sharpness of the separation). Both Xc50 and m become the characteristics of the collection efficiency. This is because the estimation of the total collection efficiency. Using the value of Xc50 alone is very difficult in a wide range of distributions of η_x.

In order to compare with the experimental results of the fractional collection efficiency, the values calculated by Equation 23 are drawn by the solid lines in Figures 18 (A) and (B) and 19 (A) and (B). From this comparison of the experimental values with those of Equation 23, this equation could describe the characteristics of the fractional collection efficiency of the experimental results especially in the region between η_x = 20 to 30% and $\eta_x = 95\%$, but below η_x = 20 to 30% and above $\eta_x = 95\%$ there was a little difference between the calculated values of Equation 23 and the experimental results.

Figures 20 and 21 show the values of Xc50 for H = 466 mm and H = 266 mm. In the case of H = 466 mm, the value of Xc50 fell within 0.4 to 6.3 μm and also in the case of η_c ≒ 99% the cut-size Xc50 ≒ 1.1 μm was reached. Then, in the case of the value of $\eta_c = 99 \pm 0.5\%$, Xc50 = 0.7 to 1.4 μm was obtained. Above V2 = 25 m/s, the values of

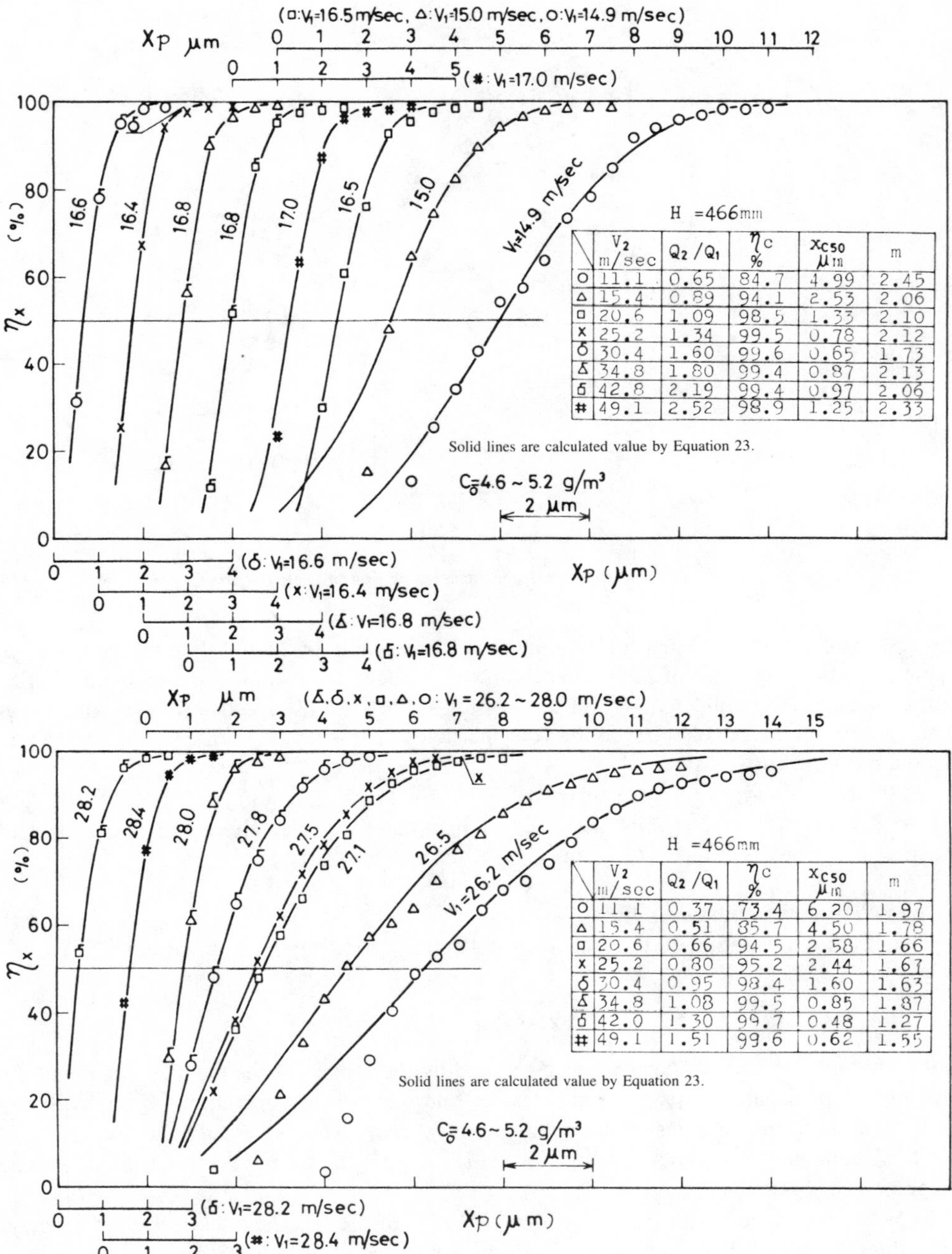

	V_2 m/sec	Q_2/Q_1	η_c %	x_{c50} μm	m
○	11.1	0.65	84.7	4.99	2.45
△	15.4	0.89	94.1	2.53	2.06
□	20.6	1.09	98.5	1.33	2.10
x	25.2	1.34	99.5	0.78	2.12
δ	30.4	1.60	99.6	0.65	1.73
Δ	34.8	1.80	99.4	0.87	2.13
б	42.8	2.19	99.4	0.97	2.06
#	49.1	2.52	98.9	1.25	2.33

	V_2 m/sec	Q_2/Q_1	η_c %	x_{c50} μm	m
○	11.1	0.37	73.4	6.20	1.97
△	15.4	0.51	85.7	4.50	1.78
□	20.6	0.66	94.5	2.58	1.66
x	25.2	0.80	95.2	2.44	1.67
δ	30.4	0.95	98.4	1.60	1.63
Δ	34.8	1.08	99.5	0.85	1.87
б	42.0	1.30	99.7	0.48	1.27
#	49.1	1.51	99.6	0.62	1.55

FIGURE 18. (Top) Fractional collection efficiency for the particle size in the case of H = 466 mm. (Bottom) Fractional collection efficiency for the particle size in the case of H = 466 mm.

Xc50 went below 3 μm. In the case of V1 = 10.1 to 23.5 m/s, even if the value of V2 was 15 to 20 m/s, Xc50 went below 3 μm. These results for the case of Q2/Q1 = 0.7 to 1.2 suggest theoretically that dust having the distribution of fly-ash as a test dust is collected above $\eta_c \geqq 90\%$. The value of Xc50 for the case of V1 = 10.1 to 11.7 m/s was coincidental with the value which was obtained by Ogawa et al. at V2 ≒ 15 m/s as shown in Figure 20, and also this tendency was very similar to that of Ogawa et al.

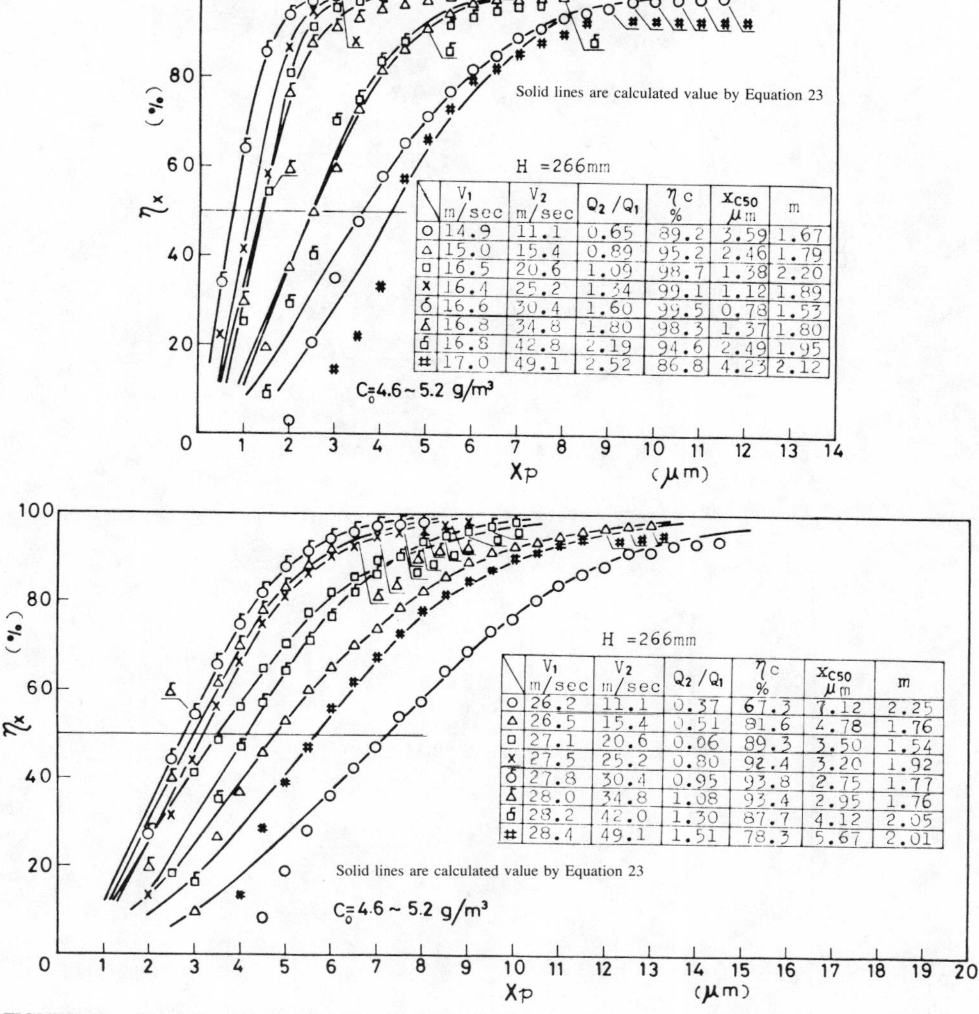

FIGURE 19. (Top) Fractional collection efficiency for the particle size in the case of H = 266 mm. (Bottom) Fractional collection efficiency for the particle size in the case of H = 266 mm.

In the case of H = 266 mm the values of Xc50 were within 1.0 to 3.5 μm for V2 ≒ 30 m/s. Also these showed the minimum values for various values of V1 as shown in Figure 21.

The values of m are shown in Figures 22 and 23 for the values of $Xc50/X_{R50}$ which are non-dimensionalized by X_{R50} = 12.44 μm. In the case of H = 466 mm for the range of $Xc50/X_{R50}$ = 0.05 to 0.2, the values of m were 2.2 to 2.8 for the condition of V1 = 10.1 to 11.7 m/s, and also m = 1.8 to 2.3 for V1 = 14.9 to 23.5 m/s, then m = 1.3 to 1.7 for V1 = 26.2 to 33.1 m/s. For the three regions of the inlet velocity, the calculated values of Rec = (Q1 + Q2)/H·ν correspond to (1.9 to 5.2) × 10^3, (2.3 to 6.3) × 10^3 and (3.4 to 7.3) × 10^3, respectively, but with an increase in values of Rec, the values of m decrease.

On the other hand, with an increasing the value of $Xc50/X_{R50}$, the values of m increase, the values of $Xc50/X_{R50}$ raise from 0.1 to 0.5, and the values of m became about 0.2 to 0.3 in each case.

In the case of H = 266 mm, for respective ranges of the values of V1, the values of m fall within the range of 1.5 to 2.5. In this case the values of Rec are (3.3 to 12.7) × 10^3,

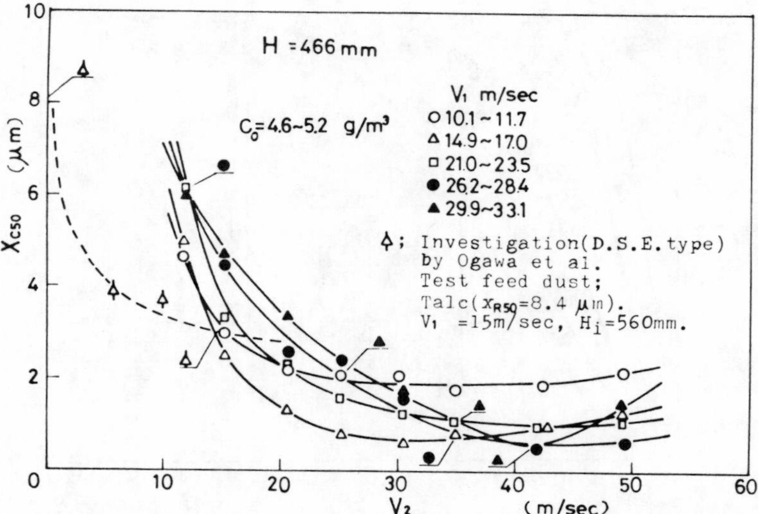

FIGURE 20. Particle size of η_x = 50% in the case of H = 466 mm.

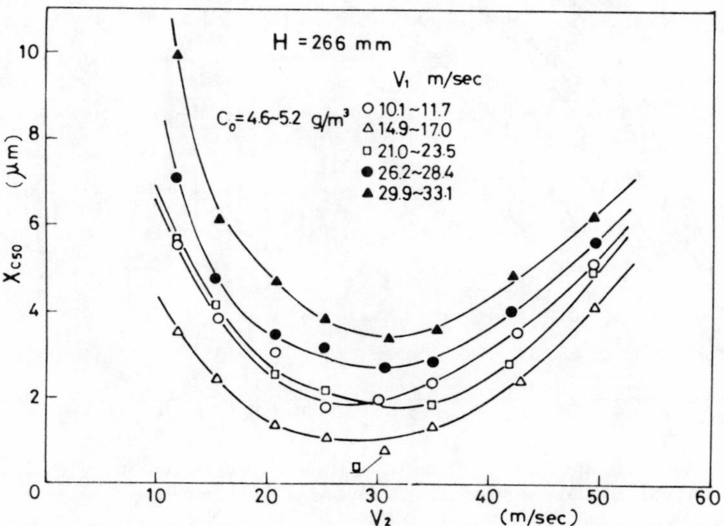

FIGURE 21. Particle size of η_x = 50% in the case of H = 266 mm.

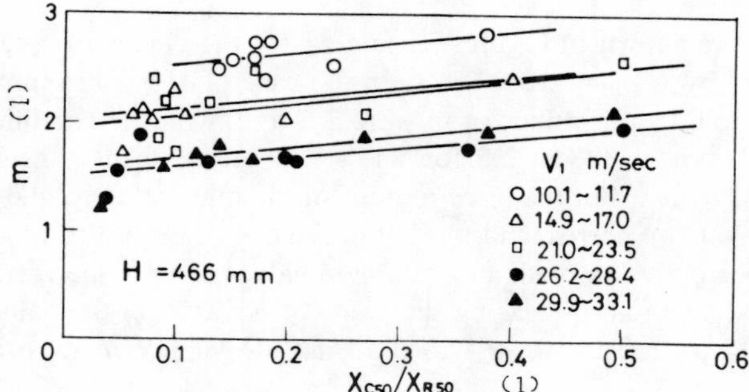

FIGURE 22. Separation index for $Xc50/X_{R50}$ in the case of H = 466 mm.

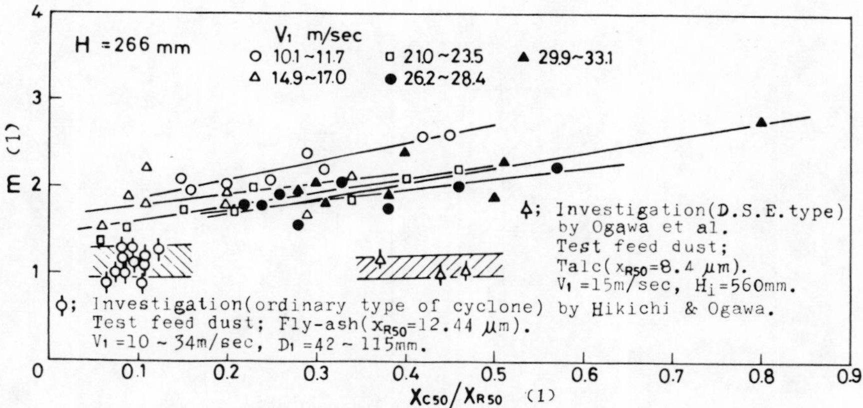

FIGURE 23. Separation index for Xc50/X_{R50} in the case of H = 266 mm.

also the values of m become smaller than the values for H = 466 mm. Moreover, the experimental values of the secondary flow type dust collector by Ogawa et al.[7,8] and of the ordinary type of cyclones by Hikichi and Ogawa are shown in Figure 23. These values of η_c correspond to 75 to 88% and 95 to 97%. Therefore, even if the values of Xc50 for each case are nearly the same, the values of η_c are different largely depending on the grade of magnitude of m.

VII. CONCLUSIONS OF THE OGAWA TYPE ROTARY FLOW DUST COLLECTOR

As the separation characteristics of a rotary flow dust collector, the total collection efficiencies η_c and the fractional collection efficiencies η_x were measured by using fly-ash as a test dust. The relationships between the values of η_c and the total pressure drops Δp_c were made clear. Then, it was made clear by experiment that the empirical equation of η_x could be expressed by Equation 23 and also that the values of Xc50 and m characterized the performances of collection efficiency. Concerning the mechanism of separation, the equi-flow rate lines and the mechanically balanced particle sizes $\overline{X}pB$ in the main body of vortex chamber were calculated. In the following these results are summarized.

1. The cut-size Xc can be estimated by Equations 16 and 17.
2. The value of η_c more than η_c = 99% can be obtained. This collector shows a higher collection efficiency than the ordinary type of cyclone. The value of Δp_c is 1.47 to 2.45 kPa and the value of E is 14.7 to 49.0 *N*–m/s.
3. The ratio Q2/Q1 which yields a high collection efficiency is limited to Q2/Q1 = 1 to 1.9.
4. Comparing the collection efficiency between H = 466 mm and H = 266 mm, in the case of H = 466 mm a higher collection efficiency is obtained and its application is wide.
5. The minimum value of Xc50 is 0.48 μm and the value of η_c is 99.7% for this case, so it is easy to obtain a value of Xc50 below 1 μm.
6. The value of m is 1.3 to 2.8, thus the value of m with the value of Xc50 is an index which shows the separation efficiency.
7. It is very difficult to infer the values of η_c and of Xc50 with a mass method from the value of $\overline{X}pB$.
8. On the pulsating dust-laden gas flow, the collection efficiency η_c decreases by a small amount.[9]

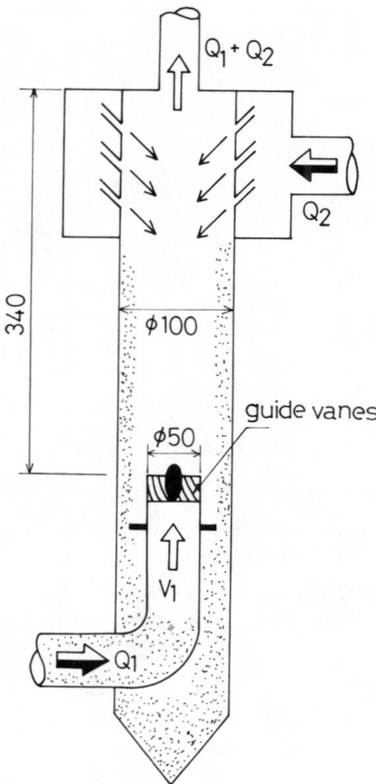

FIGURE 24. One type of rotary flow dust collector.

VIII. LANCASTER AND CILIBERTI EXPERIMENT

Lancaster and Ciliberti (1976)[10] investigated the separation performance of rotary flow dust collectors, as shown in Figure 24. In this type of collector, main primary flow Q1 and the secondary flow Q2 together flow out through the inlet pipe (exhaust pipe). The test dust was the lime-powder ($R(Xp) = \exp(-0.118 \cdot Xp^{1.4})$) of density $\rho_p = 2$ g/cm^3 and of the mean diameter $X_{R50} = 3.7$ μm. From those experimental results, one of the most important characteristics was the ratio of flow rates Q1 to Q2 which is related to the separation performance η_c, namely, $\eta_c = 60$ to 66% for Q2/Q1 = 0.58, $\eta_c = 77$ to 83% for Q2/Q1 = 1.7, as shown in Figure 25. These results show the very similar characteristics to the Ogawa and Hikichi experiments.

IX. FLOW SYSTEMS

In the development of the rotary flow dust collectors in Siemens A.G., three types of the flow systems[6] were constructed, as shown in Figure 26.

A. Type I

The main primary flow Q1 (dust-laden gas flow) becomes the turbulent rotational flow for the upward direction passing through the guide vanes, while the secondary flow becomes the rotating flow passing through the guide vanes for the downward direction along the surface of the outer pipe of a diameter D1. This secondary flow Q2 intensifies the primary vortex flow and transports the separated dust to the dust bunker. Then the primary flow Q1 and the secondary flow Q2 together flow out through the inlet pipe (exit pipe), for example, see the Lancaster and Ciliberti experiment (1976).

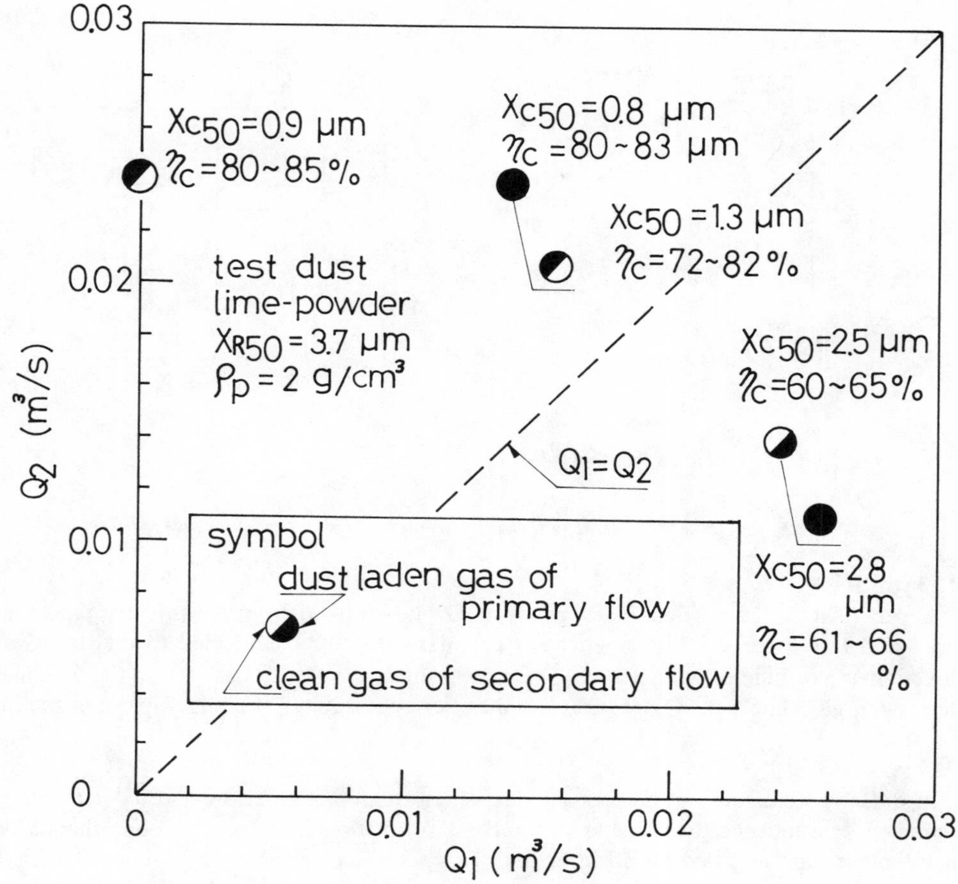

FIGURE 25. Relationship between Q1 and Q2 for the collection efficiency and cut-size.

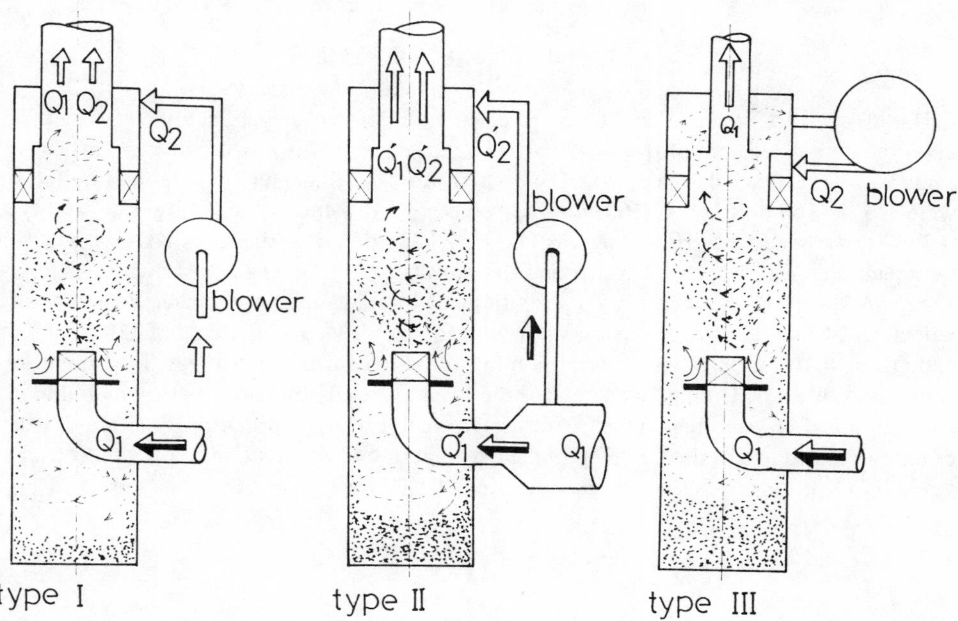

FIGURE 26. Three types of flow systems of rotary flow dust collectors.

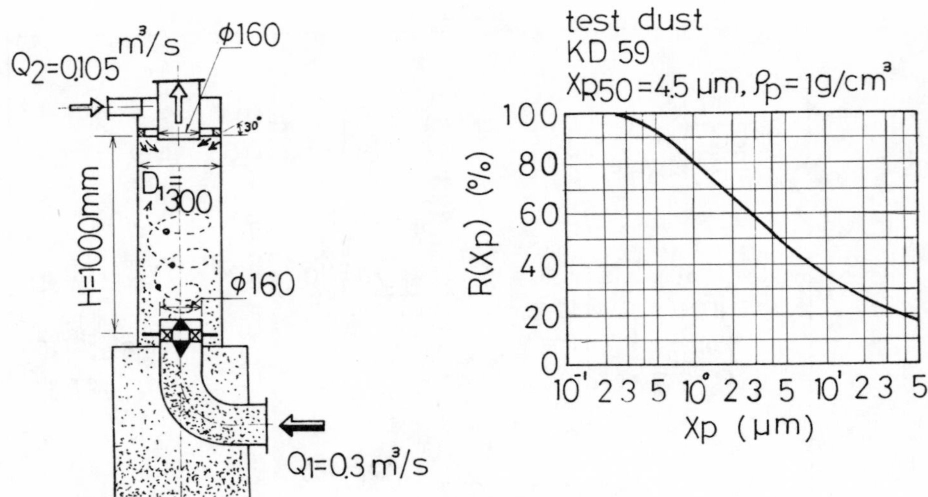

FIGURE 27. Particle size distribution and the main sizes of the rotary flow dust collector.

B. Type II

A part of the primary dust-laden gas flow Q1 is absorbed from the primary pipe (inlet pipe) and this absorbed dust laden gas (secondary flow) Q2́ is ejected to the downward direction as a turbulent rotational jet. There the primary dust-laden gas Q1́ and the secondary dust laden gas Q2́(Q1́ + Q2́ = Q1) together flow out through the inlet pipe (exit pipe).

C. Type III

In order to make strong the primary dust-laden rotational gas flow and to transport the separated dust into the dust-bunker, the secondary rotational flow ejects along the surface of the outer pipe in a downward direction. This secondary gas flow re-circulates again to the secondary jet nozzles passing through the blower, for example see the Hikichi and Ogawa experiment (1979).

X. BUDINSKY EXPERIMENT

Budinsky (1972)[11] made an attempt to calculate the fractional collection efficiency based upon the calculated loci of the solid particles in the rotary flow dust collector, as shown in Figure 27. The main diameter was D1 = 300 mm, the diameters of inlet and outlet pipes were Do = 160 and D2 = 160 mm, respectively. Also the primary flow rate was Q1 = 0.3 m³/s, the secondary flow rate was Q2 = 0.105 m³/s. Test dust was KD-59, which had the mean diameter X_{R50} = 4.5 μm and the density ρ_p = 1 g/cm³, as shown in Figure 27.

Figure 28 shows the measured distributions of the tangential Vθ, radial Vr and axial Vz velocities of the gas flow by using Pitot-tube for the driving conditions of Q1 = 0.3 m³/s and Q2 = 0.105 m³/s. Then, based upon the velocity distributions of gas flow in the rotary flow dust collector, Budinsky calculated the loci of the DK-59 particles of which diameters Xp = 1.2 and 10 μm obeyed the Stokes drag force, as shown in Figure 28. From the results of the calculated loci of the DK-59 particles, he determined the fractional collection efficiency curve η_x (Xp) corresponding to cut-size Xc = 1.7 μm (η_x = 50%), as shown in Figure 29.

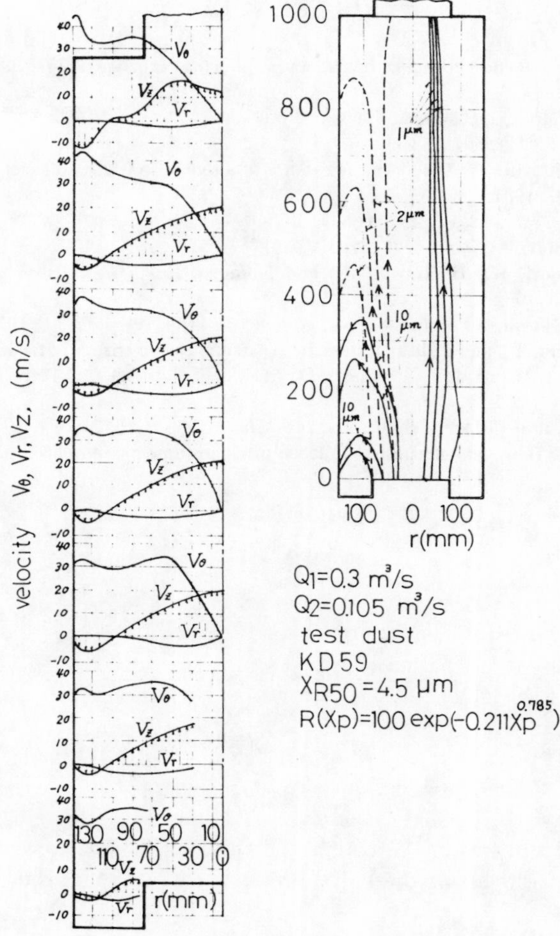

FIGURE 28. Velocity distributions of air flow and the calculated results of the loci of solid particles.

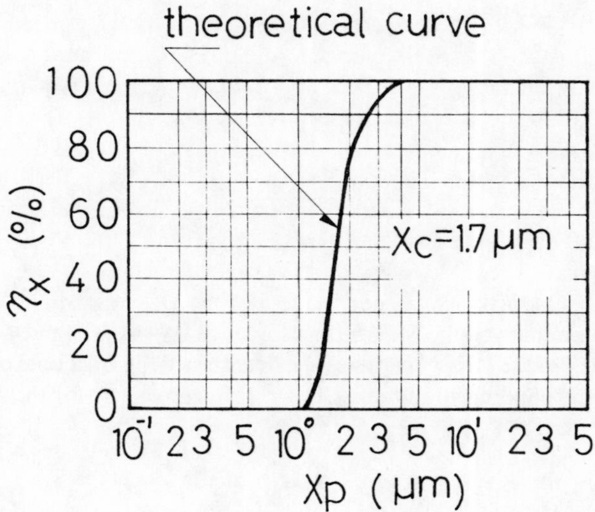

FIGURE 29. Fractional collection efficiency curves determined by the calculations.

REFERENCES

1. **Pieper, R.,** The tornado dust collector. Fourteen years practical experience, *VDI-Berichte (VDI-Rep.),* 294, 93, 1977.
2. **Schmidt, K. R.,** *Staub,* 23(11), 491, 1963.
3. **Klein, H.,** *Staub,* 23(11), 501, 1963.
4. **Ogawa, A. and Hikichi, T.,** Theory of the cut-size of a rotary flow dust collector, *Bull. Jpn. Soc. Mech. Eng.,* 24(188), 340, 1981.
5. **Klein, H.,** *Energ. Tech.,* 18(6), 228, 1966.
6. **Alt, C. and Schmidt, P.,** *Staub,* 29(7), 263, 1969.
7. **Ogawa, A., Ikemori, K., Hirasawa, M., and Komuro, K.,** *J. Res. Assoc. Powder Tech. Jpn.* (in Japanese), 9(6), 17, 1972.
8. **Hikichi, T. and Ogawa, A.,** *J. Jpn. Soc. Air Pollut.,* (in Japanese), 12(3), 1, 1977.
9. **Ogawa, A., Hikichi, T., and Fujita, Y.,** On the separation characteristics of the rotary flow dust collector for the steady and pulsating dust laden gas flow, 5th Congr. Int. Del Aire Puro, Buenos Aires, October, 1980.
10. **Lancaster, B. W. and Ciliberti, D. F.,** *Am. Inst. Chem. Eng. J.,* 22(2), 394, 1976.
11. **Budinsky, K.,** Die Bewegung der festen Teilchen im Drehströmungsentstauber, *Staub,* 32(3), 87, 1972.

Chapter 3

BAG FILTERS

I. INTRODUCTION

Up to this time, there have been many theories and experimental work done on bag filters concerning pressure drop, separation mechanism with the fluid flow pattern, and the collection efficiency. In this chapter the author wants to describe mainly the fundamental theories and characteristics of bag filters used in general industry.

Figure 1 shows the relationship between pressure drop and filter thickness, where W(Pa/(l/min·cm^2)) means the pressure drop per gas flow rate (l/min·cm^2) into the filter and H(cm/(g/cm^2)) means the specific thickness (cm) of the filter per 1 g filter mass per filter area (cm^2). From this figure by Kaufmann (1936), we find that the pressure drop strongly depends on the physical and chemical properties of the filter materials.[1] Figure 2 shows a relationship between the pressure drop and the collection efficiency η_c. In this experiment, the diameter of the test dust (fly-ash) was smaller than Xp = 1 μm.

II. EARLY THEORY OF FIBROUS FILTERS

Now C(1) represents the packing density of fiber solidity which is the ratio of the volume of all of the fibers to the volume of the filter.[2] Supposing that L is the length of all of the fibers of radius R in the unit thickness to the air flow of the unit cross-flow area, as shown in Figure 3, then the packing density or solidity can be represented as

$$C = \pi \cdot R^2 \cdot L = 1 - Po \tag{1}$$

where Po(1) means the porosity of the fibrous filter. When the filter is composed of fibers, length Li and radius Ri, then the solidity C can be represented as

$$C = \pi \cdot \Sigma\, Ri^2 \cdot Li \tag{2}$$

Suppose that the collision efficiency of the aerosol particles, or of the solid particles with a fiber of radius R, is η_i, which depend on both particle and fiber parameters. Assuming that the air flow through the filter may be nearly always in a laminar state, hence η_i, which is defined as Y_i/R, is equal to the ratio of the distance between two limiting stream lines of flow approaching a fiber of the diameter 2R. The limiting stream lines are such that all of the particles (aerosols, fumes) between them will strike the fiber, whereas no particles outside the limits will encounter it, as shown in Figure 4, where the single fiber efficiency is η_i, provided that the particles which strike on the fiber remain adhered to it.

Then consider a small filter element dX with a unit cross flow area at a distance X from the entrance face, as shown in Figure 5, the length of fiber in the small element as L·dX. When the particles entering into this element contain n (particles/m^3) particles per unit volume of air and also the mean velocity of air in the filter is $\overline{V}$(m/s), then the number of particles entering per unit time is n·$\overline{V}$. The area of the fiber capable of representation in this small element is

$$L \cdot 2R \cdot dX \tag{3}$$

and also the effective filtering area can be represented as

$$\eta_i \cdot L \cdot 2R \cdot dX \tag{4}$$

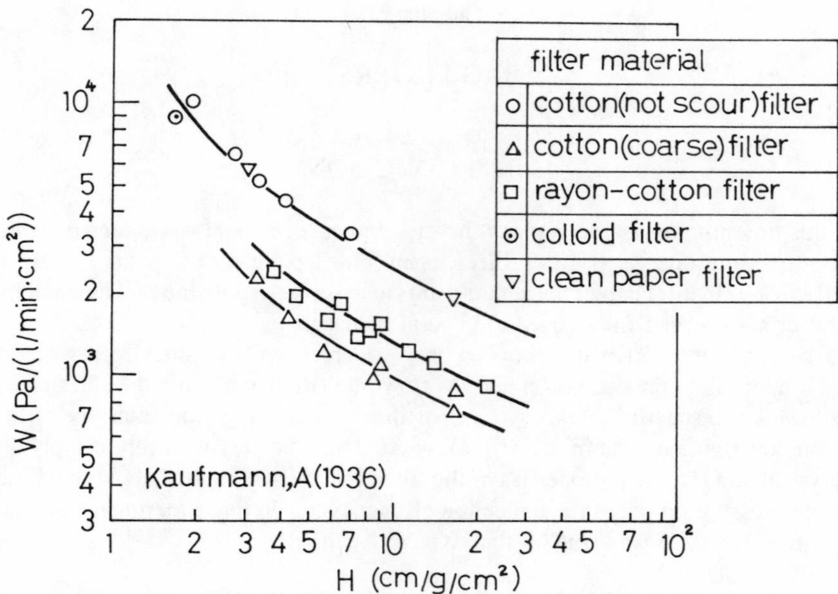

FIGURE 1. Relationship between the pressure drop and the filter thickness.

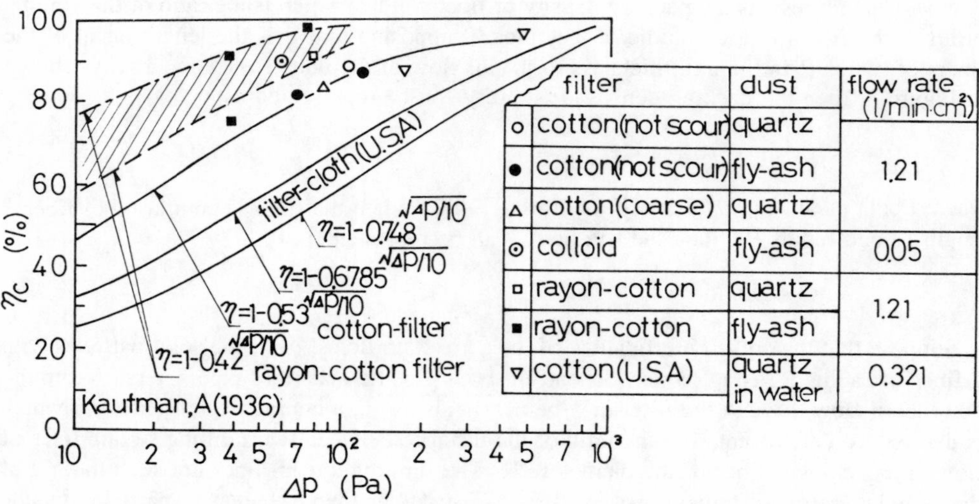

FIGURE 2. Collection efficiency and pressure drop for the particles of diameter Xp = 1 μm and smaller.

Therefore, the rate of removal of particles by the small element of filter per unit cross flow area becomes

$$n \cdot \overline{V} \cdot \eta_i \cdot L \cdot 2R \cdot dX \tag{5}$$

Here, a relationship between the gas velocity $\overline{V}$ inside the filter and the face velocity $\overline{V}o$, or approaching velocity $\overline{V}o$, to the filter can be written

$$\overline{V} = \frac{Vo}{(1 - C)} \tag{6}$$

where $\overline{V}o$ is defined as

$$\overline{V}o = \frac{Qo}{A} \tag{7}$$

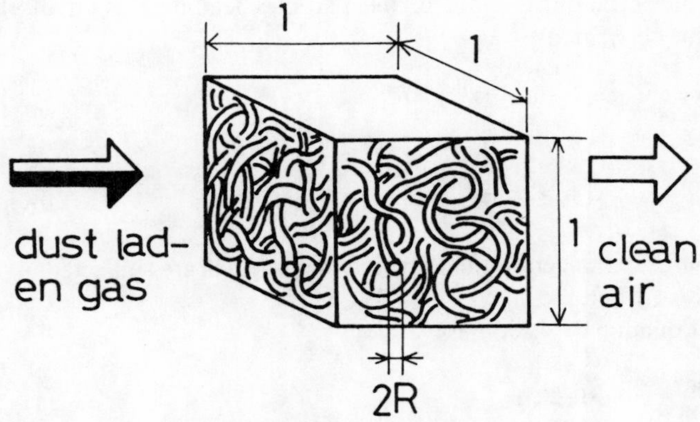

FIGURE 3. Illustration of the early theory of the fibrous filter.

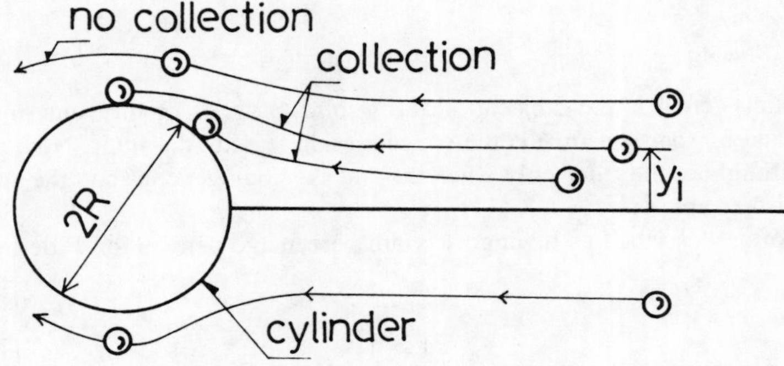

FIGURE 4. Definition of a single fiber efficiency.

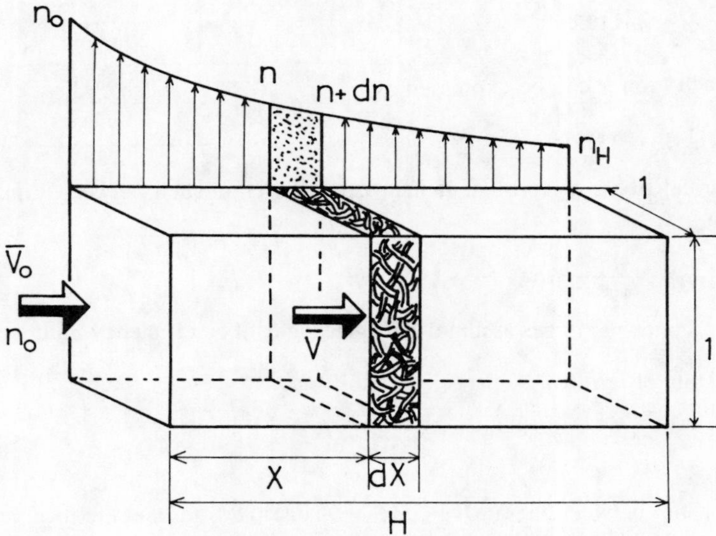

FIGURE 5. Small element dX of a filter of thickness H.

Qo(m^3/s) is gas flow rate and A(m^2) is the cross flow area of the filter. Then the number of particles leaving this small area per unit time at the distance X + dX can be written

$$n \cdot \overline{V}(1 - \eta_i \cdot L \cdot 2R \cdot dX) \qquad (8)$$

Therefore, the concentration n + dn of the particles leaving this small element per unit volume of gas can be written

$$n + dn = n(1 - \eta_i \cdot L \cdot 2R \cdot dX) \quad (9)$$

so that

$$dn = -\eta_i \cdot L \cdot 2R \cdot n \cdot dX \quad (10)$$

represents the change of concentration to which the particles are subjected in passing through the small element of filter.

Consequently Equation 10 becomes

$$\frac{dn}{dX} = -\eta_i \cdot L \cdot 2R \cdot n \quad (11)$$

Integrating Equation 1 across the thickness H of the filter, we can obtain

$$\frac{n_H}{no} = \exp(-\eta_i \cdot L \cdot 2R \cdot H) \quad (12)$$

where no(particles/cm^3) is the concentration of solid particles or dust entering a filter of thickness H, and n_H (particles/m^3) is the concentration leaving the filter. Note that η_i is the same value throughout the filter only when the filter is homogeneous and the solid particles or dust are all the same size.

Filters are often described as having a certain percentage penetration P defined as

$$P = \frac{n_H}{no} \times 100 \quad (13)$$

Also the value ($\eta_i \cdot L \cdot 2R$) (1/m) per unit thickness of filter is called the filtration index γ defined as

$$\gamma = \eta_i \cdot L \cdot 2R \quad (14)$$

Therefore, penetration P can be expressed as

$$P = 100 \exp(-\gamma H) \quad (15)$$

For the cylindrical fibers of constant radius, L can be eliminated between Equation 1 and Equation 14 as

$$\gamma = \frac{2 \cdot \eta_i \cdot C}{\pi \cdot R} \ (1/m) \quad (16)$$

Now on the contrary, if we assume that the single fiber efficiency η_i depends on the distance X as

$$\eta_{i1} = \eta_i \left(1 - \frac{X}{2 \cdot H}\right) \quad (17)$$

$$\eta_{i2} = \eta_i \left\{1 - \frac{1}{2}\left(\frac{X}{H}\right)^2\right\} \quad (18)$$

as shown in Figure 6, then we can obtain the equations of penetration as

$$P1 = 100 \exp\left(-\frac{3}{4}\gamma H\right) \quad \text{for } \eta_{i1} \quad (19)$$

$$P2 = 100 \exp\left(-\frac{5}{6}\gamma H\right) \quad \text{for } \eta_{i2} \quad (20)$$

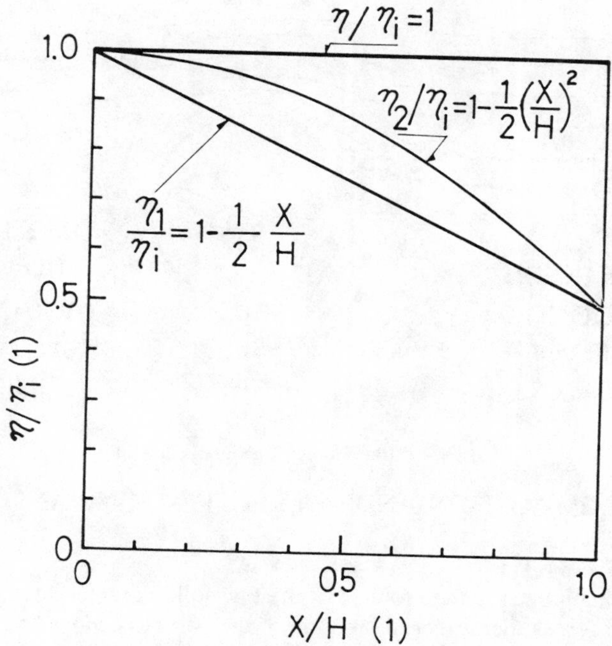

FIGURE 6. Characteristics of the collision efficiencies.

Table 1
RELATIONSHIP BETWEEN γH AND PENETRATION (%)

γH	P Equation 15	P1 Equation 19	P2 Equation 20
13.9	0.000092	0.00297	0.000932
12.2	0.000503	0.0106	0.00384
11.5	0.00101	0.0180	0.00689
9.9	0.00502	0.0596	0.0261
9.2	0.0101	0.1008	0.0468
7.6	0.0500	0.335	0.178
6.9	0.10	0.566	0.318
5.3	0.50	1.88	1.207
4.6	1.0	3.17	2.16
3.0	5.0	10.54	8.21
2.3	10.0	17.82	14.71
1.61	20.0	29.89	26.16
0.694	50.0	59.42	56.08
0.511	70.0	68.16	65.32
0.223	80.0	84.60	83.04
0.105	90.0	92.43	91.62

Those numerical values are written in Table 1. From Table 1, the collection efficiency η_c strongly depends on the characteristics of the efficiency η_i. Here the collection efficiency η_c can be calculated from Equation 15 as

$$\eta_c = \frac{no - n_H}{no} = 100 - P\ (\%) \tag{21}$$

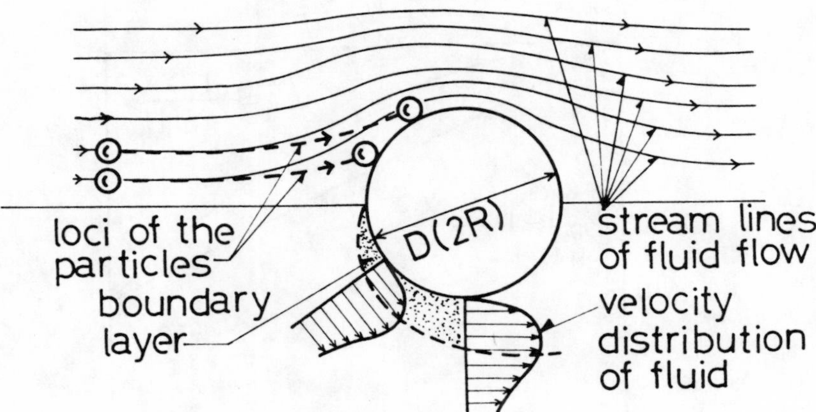

FIGURE 7. Illustration of the inertia deposition.

III. IMPACTION AND COLLECTION MECHANISMS

A. Collection Mechanisms

A fundamental approach to the problem of the fine solid particle, dust, or fume collection is to estimate the forces that affect the motion of the particle and cause it to move from an aerosol or particle stream, onto a collecting surface.[3]

The important dimensionless parameters which are essentially recognized as a comparison of the various forces tending to cause impaction with the fluid resistance force opposing particle motion, or as a comparison of two dimensions of the physical system, characterize the mechanisms of collection and can be defined as follows.

1. Inertia Deposition

For collection by the inertia deposition, as shown in Figure 7, solid particles or fumes have enough inertial force that they cannot follow the stream lines around the fiber and then impact on its surface. This collection mechanism involves a collision of the particle on the barrier. Therefore, the efficiency η_i of the inertia deposition depends on the Stokes number defined as

$$St = \Psi = \frac{\rho \cdot Xp^2 \cdot \overline{V}o}{18 \cdot \eta \cdot D} \tag{22}$$

where D is a representative target diameter and $\overline{V}o$ is an approach velocity of a particle.

2. Interception

A second collection mechanism is the particle interception. In this collection mechanism, solid particles or fumes have less inertial force and can barely follow the stream lines around the fiber or the cylinder. Even if it does not actually touch the fiber, the particles are almost completely immersed in the viscous stream around the barrier which will be enough to slow it down so it will graze the barrier and stop on the surface of the fiber or cylinder, as shown in Figure 8.

When this situation occurs, the porosity of the fabric filter medium decreases and the true filtering surface begins to form. Pore bridging can be completed rapidly if the ratio of pore diameter D to particle diameter Xp is less than 10. If that ratio is larger than 10, a considerable amount of seepage can occur. Torgeson (1961)[4] expressed the efficiency η_{int} of the collection by the interception for the Stokes law region as

$$\eta_{int} = 0.00759\, C_D \cdot Rep \cdot \left(\frac{Xp}{D}\right)^{1.5} \tag{23}$$

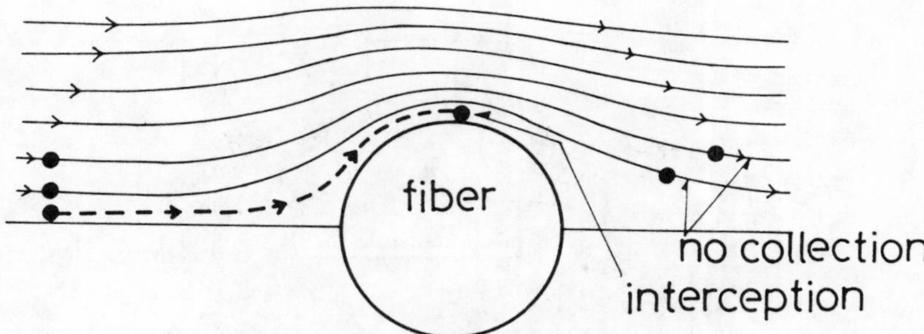

FIGURE 8. Illustration of the interception.

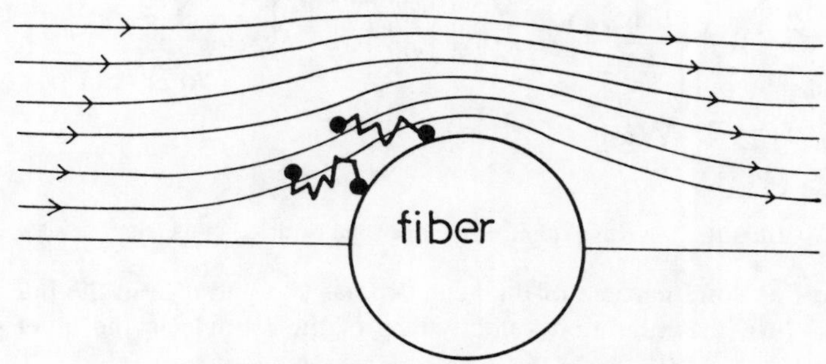

FIGURE 9. Illustration of the diffusion.

where C_D is the drag coefficient of the barrier or target and Rep is Reynolds number based on the particle diameter Xp.

The maximum efficiency of the collection which can be achieved for any flow region is

$$\eta_{int}(\text{Max}) = \frac{1 + Xp}{D} \tag{24}$$

Then the mechanisms of the inertia deposition and interception usually account for 99.9% of the collection of particles larger than Xp = 1 μm.

3. Diffusion

The third collection mechanism is very important for fine solid particles or fumes that are below 1 μm in aerodynamic diameter. In this situation, fine solid particles or fumes are so small that their individual motions can be affected by collisions on the molecular or atomic level. Collection of these fine particles is a result of Brownian random motion. When the particle size is less than 0.1 μm, Brownian diffusion becomes significant, as shown in Figure 9.

The collection efficiency η_d by diffusion has been described by Torgenson as

$$\eta_d = 0.75\left(\frac{C\ Rep}{2}\right)^{0.04} \cdot (\bar{V}o\ D)^{-0.6} \cdot \frac{3 \cdot \pi \cdot \eta \cdot Xp}{K \cdot C} \tag{25}$$

where C is Cunningham slip correction factor and K is Boltzmann constant $K \doteq 1.38 \times 10^{-23}$ J/K.

4. Electrostatic Attraction

When solid particles or fumes are charged by the corona current (or by some other way), the particles may be encouraged to impact upon the collecting surface by an actual or induced electrostatic force between the particle and the collector.

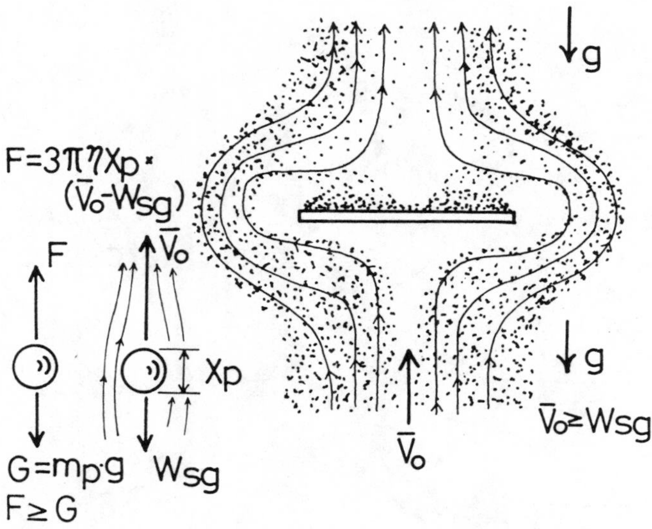

FIGURE 10. Gravity settling of the fine solid particles on the plate.

Then when the fine solid particle of diameter Xp has charged q(c) in the intensity of the electric field E(kV/m), the migration velocity $\bar{U}p$ of the particle in the quiet gas can be written

$$\bar{U}p = \frac{q \cdot E}{3 \cdot \pi \cdot \eta \cdot Xp} \ (m/s) \tag{26}$$

A detailed description of this electrostatic attraction is presented in Chapter 4, Electrostatic Precipitators.

5. *Gravity Settling*

When the fine solid particles of diameter Xp follow the stream lines of the upward gas flow around a plate against gravity force, as shown in Figure 10, the fine solid particles settle down on the surface of the plate behind the plane approaching the flow. Here Wsg is the terminal velocity in gas defined as

$$Wsg = \frac{\rho_p \cdot g \cdot Xp^2}{18 \cdot \eta} \tag{27}$$

and also the force F acting on the solid particle can be written

$$F = 3 \cdot \pi \cdot \eta \cdot Xp(\bar{V}o - Wsg)$$

Figure 11 represents the calculated results by Kaufmann (1936)[1] of deposition in real wool filters due to interception, inertia, and Brownian motion. In this figure, the mean diameter of the wool filter fiber might be 2R = 20 μm and the mean air velocity through it is $\bar{V}$ = 15 cm/s, therefore the flow Reynolds number around this fiber becomes

$$Rep = \frac{2R \cdot \bar{V}}{\upsilon} = 0.2$$

where the kinematic viscosity υ of air is $\upsilon \fallingdotseq 0.15$ cm²/s. Consequently, the air flows around the fiber, as this low Reynolds number is viscous and is unaffected by the inertia of the air.

A new theory combining diffusion, interception, and inertia of the particle was derived by Davies (1952). The filter was described in Equations 1 to 16 applying the concept of a single fiber efficiency as

$$\eta_i = \frac{y}{R} + \frac{Wsg}{\pi \cdot \bar{V}o} \tag{28}$$

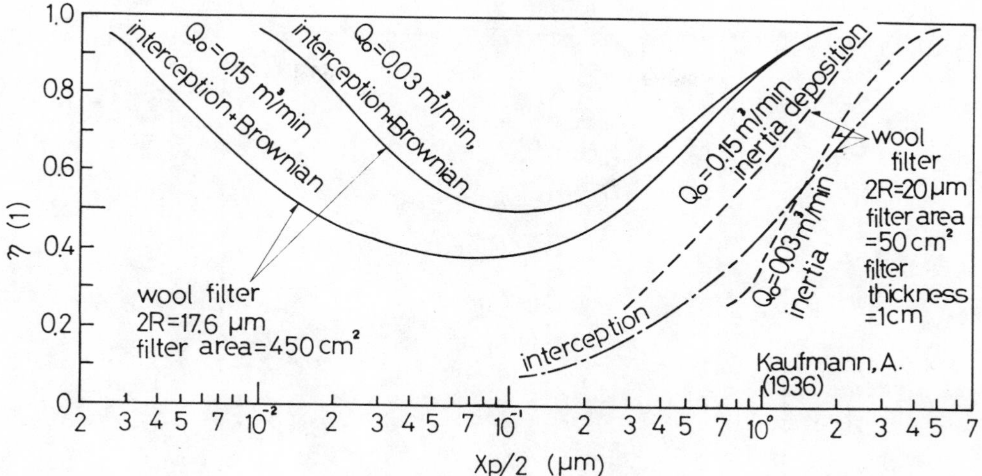

FIGURE 11. Calculation of deposition in real wool filter due to interception, inertia, and Brownian motion.

In this equation, the second term means the effect of deposition by the particles settling under the force of gravity upon the fibers with the terminal velocity Wsg.

The three mechanisms, diffusion, interception, and inertia were combined in a single expression as follows:

$$\frac{y}{R} = \left\{\frac{a}{R} + \left(0.25 + 0.4\cdot\frac{Xp}{2R}\right) Pa - 0.0263\cdot\frac{Xp}{2R}\cdot Pa^2\right\} (0.16 + 10.9C_k - 17\cdot C_k^2) \tag{29}$$

where the parameter Pa is the sum of the Stokes number St and a diffusion number $Dp/\overline{V}\cdot R$ which is twice the reciprocal of the diffusion analogue of the Peclet number of heat transfer. Langmuir's approximation for Dp might be applied as

$$Dp = \frac{2 \times 10^{-16}}{(Xp/2)^2}\ (cm^2/s) \tag{30}$$

The value of Dp in this approximation are good within 16% for $0.05 \leqq Xp/2 \leqq 0.1$ μm at R = 0.15 μm. Dp is 28% low and 40% at R = 0.2 μm. Dp is 43% high at R = 0.01 μm. Hence the value of Pa to be used can be written

$$Pa = \frac{\rho\cdot Xp^2\cdot\overline{V}}{18\cdot\eta\cdot R} + \frac{8 \times 10^{-16}}{Xp^2\cdot\overline{V}\cdot R} \tag{31}$$

The term containing Ck, which allows for the mutual influence of the fibers, came from an exact hydrodynamical solution for flow through a grid by Kovasznay (1948)[5] which was applied as a basis for extrapolation from the viscous flow curve of interception at Ck = 0 and P = 0 for $Ck \leqq 0.213$.

Then differentiating Equation 31 with respect to velocity $\overline{V}$ and keeping Xp constant gives the condition for the maximum penetration when the velocity is varied as

$$\overline{V} = \left\{10^{9.4}(Xp/2)^2\right\}^{-1} = (10^{9.4}\cdot Xp^2/4)^{-1} \tag{32}$$

In general the most penetrating particle size Xp decreases rapidly when either the fiber radius or the air velocity is increased, thus confirming, as indicated above, that diffusion is of little importance in filters of coarse fibers ($\geqq$ 5 μm radius) and high velocity loading ($\geqq$ 5 cm/s).

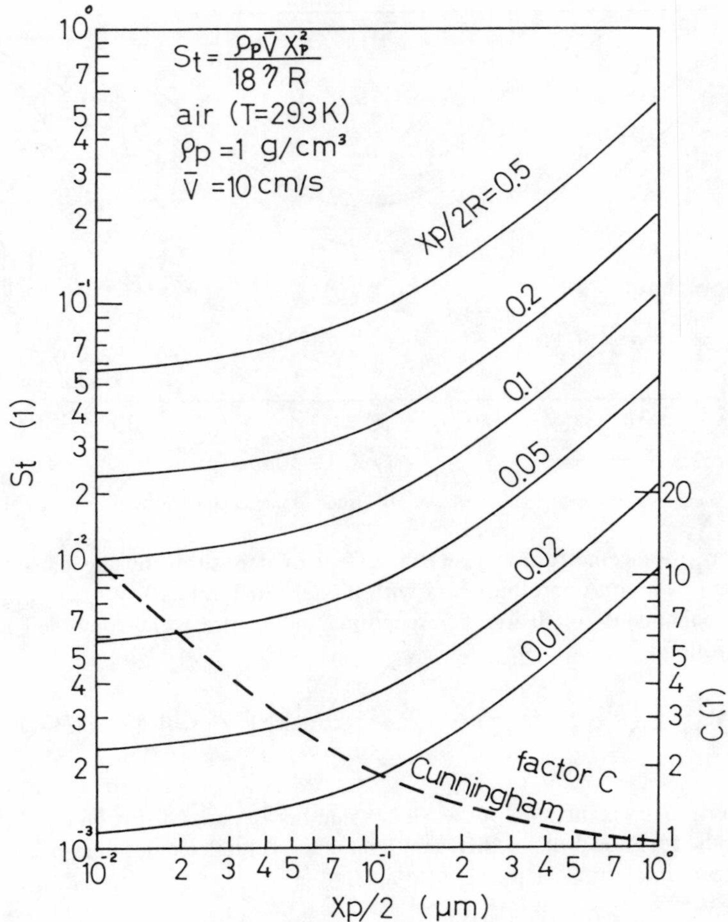

FIGURE 12. Relationship between Stokes number St and particle diameter Xp.

On the other hand, Yoshioka, Emi, and Yasunami (1969)[6] combined all the preceding results and added diffusive deposition according to the formula of Friedlander[7] for an isolated cylinder

$$\eta_d = 6 \cdot Rep^{1/6} \cdot Pe^{-2/3} \tag{33}$$

and they also gave an approximation for interception as

$$\eta_i = 3 \cdot Rep^{1/2} \cdot \left(\frac{Xp}{2R}\right)^2 \tag{34}$$

Here, in order to estimate a relationship between the Stokes number and particle diameter Xp of unit density in air which is depended on a parameter Xp/2R, the author introduced Figure 12, where air flow velocity in the filter is $\overline{V}$ = 10 cm/s constant and air temperature is T = 293 K.

B. Mathematical Description of the Inertia Impaction

Here, in order to illustrate the mathematical description of the inertia impaction of the particle on the rectangular infinite long obstacle, as shown in Figures 13 and 14, we apply

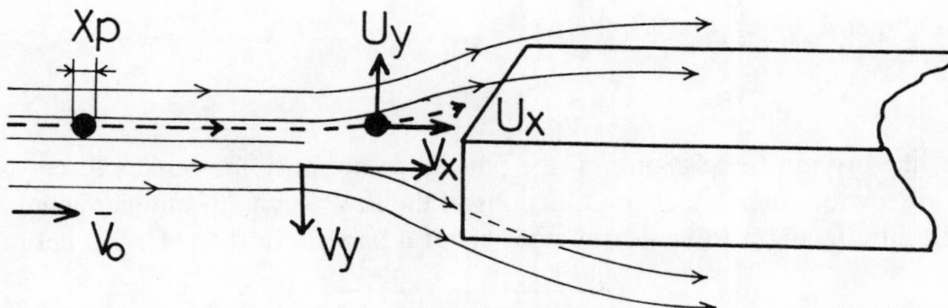

FIGURE 13. Illustration of the mathematical description of the inertia impaction of the solid particle.

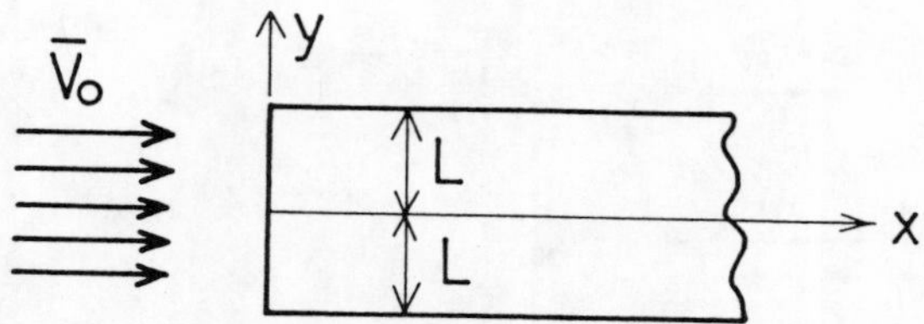

FIGURE 14. Illustration of the rectangular infinite long obstacle.

a paper of Lewis and Brum written in 1956.[8] Before deriving the differential equations of the motion of the particle in gas, we must assume the following conditions:

1. The particle rests at the infinite distance in front of the rectangular obstacle in the gas stream
2. The driving force acting on the particle is the Stokes drag force which is related to the velocity difference between particle and fluid
3. The particle is a spherical particle of the diameter Xp

Now, denoting that the absolute velocities of particles for x and y directions are Ux and Uy, and that of fluid are Vx and Vy, respectively, then the equation of motion of the particle can be described for x and y directions as

$$m_p \frac{dUx}{dt} = 3 \cdot \pi \cdot \eta \cdot Xp(Vx - Ux) \tag{35}$$

$$m_p \frac{dUy}{dt} = 3 \cdot \pi \cdot \eta \cdot Xp(Vy - Uy) \tag{36}$$

where mp is particle's mass of density ρ_p. Here, in order to transform Equations 35 and 36 to the dimensionless equation, using the free stream velocity $\overline{V}o$ of gas in front of the obstacle as

$$Ux = \overline{V}o \cdot u_x, \; Uy = \overline{V}o \cdot u_y, \; Vx = \overline{V}o \cdot v_x, \text{ and } Vy = \overline{V}o \cdot v_y \tag{37}$$

and substituting those defined quantities into Equations 35 and 36, we can obtain the following equations

$$\frac{du_x}{d\tau} = \frac{18 \cdot \eta \cdot L}{\rho_p \cdot \overline{V}o \cdot Xp^2} \cdot (v_x - u_x) = \frac{1}{K}(v_x - u_x) \tag{38}$$

$$\frac{du_y}{d\tau} = \frac{18 \cdot \eta \cdot L}{\rho_p \cdot \overline{V}o \cdot Xp^2} \cdot (v_y - u_y) = \frac{1}{K}(v_y - u_y) \tag{39}$$

where the dimensionless time τ is defined as

$$t = \frac{L}{\overline{V}o} \cdot \tau \tag{40}$$

However, the driving force acting on the particle is not only the Stokes force, but can be an Allen or Newton force which are functions of the flow Reynolds number about a particle. Therefore, introducing a Reynolds number about a particle of diameter Xp defined as

$$Rep = \frac{\rho \cdot \overline{V}o \cdot Xp}{\eta} \sqrt{(v_x - u_x)^2 + (v_y - u_y)^2} \tag{41}$$

so the above equation of motion can be transformed as

$$\frac{du_x}{d\tau} = \frac{C_D \cdot Rep}{24} \cdot \frac{1}{K}(v_x - u_x) \tag{42}$$

$$\frac{du_y}{d\tau} = \frac{C_D \cdot Rep}{24} \cdot \frac{1}{K}(v_y - u_y) \tag{43}$$

where C_D is the drag coefficient. We define a Reynolds number Reo about a particle which is located enough distance in front of an obstacle as

$$Reo = \frac{\rho \cdot \overline{V}o \cdot Xp}{\eta} \tag{44}$$

Because the drag coefficient C_D depends on the particle velocity and its location at each moment, it is very difficult to directly solve the differential equations of particle motion. Therefore, Lewis and Brun applied the differential analyzer. In order to make use of this analyzer for calculating the locus of the particle, the following procedures are necessary:

1. The values of C_D, which are a function of Reynolds number Rep are known
2. The values of v_x and v_y for the velocity component of gas flow must be known as a function of the location in front of the obstacle.

Here, in order to determine the velocity components of a gas flow, the Schwarz-Christoffel method is applied.[9] The results of this method are as follows:

$$v_x = \frac{\sinh p \cdot \cosh p - \sinh p \cdot \cos q}{\sin^2 q + \sinh^2 p} \tag{45}$$

$$v_y = \frac{\cosh p \cdot \sin q - \sin q \cdot \cos q}{\sin^2 q + \sinh^2 p} \tag{46}$$

$$X = -\frac{1}{\pi} \cdot (p + \sinh p \cdot \cos q) \tag{47}$$

$$Y = \frac{1}{\pi} \cdot (q + \cosh p \cdot \sin q) \tag{48}$$

where the values of p and q are arbitrary parameters which determine the velocity components v_x and v_y are also the positions X and Y in the flow field.

Here, assuming that the particle starts at the location of $y = y_o$ in front of the obstacle and reaches to $y = y_s$ on the obstacle, then a relationship between y_o and y_s can be written

$$yo = f(y_s) \tag{49}$$

Now the values of f(0) and f(1) show the values of y_s to be $y_s = 0$ and $y_s = 1$, therefore we can use the assumed values y_o. Figure 15 shows a numerical example of the loci of the

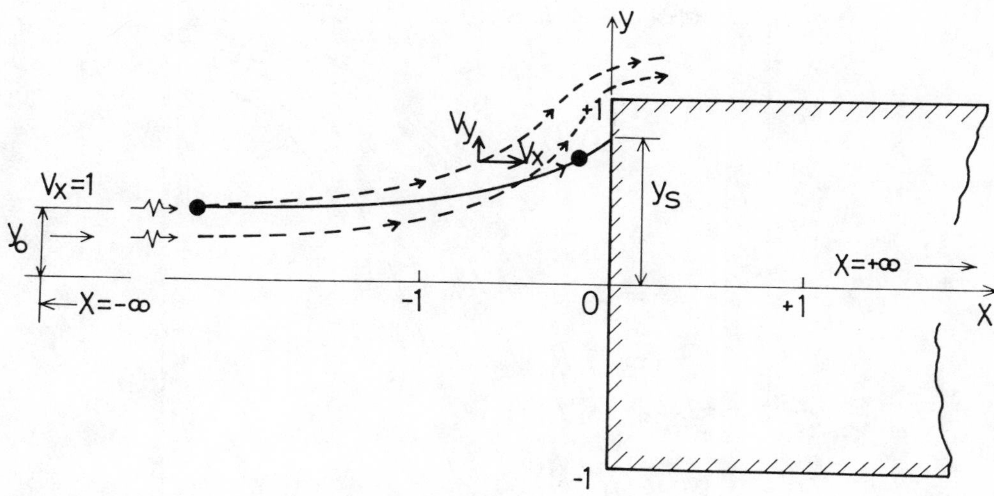

FIGURE 15. Numerical examples of the loci of the stream line of gas flow and a locus of the water droplet.

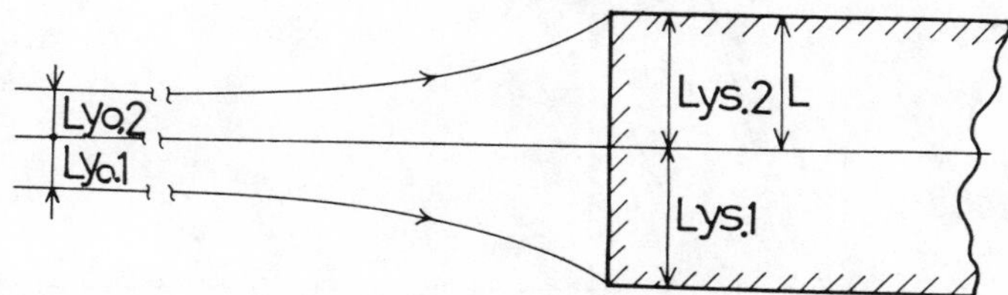

FIGURE 16. Illustration of the loci of the water droplets to the rectangular obstacle.

stream line of gas flow and a locus of the particle of water droplet which starts at $y = y_o$ and reaches $y = y_s$ on the obstacle.

Here, assuming that water droplets of the same size Xp move to the rectangular obstacle of the width 2L and that the starting points are Lyo.1 and Lyo.2, as shown in Figure 16, then the total flowing quantity of the water droplets per unit length of span can be written

$$G = \bar{V}o \cdot Wo \cdot L\,(Yo.2 - Yo.1) \tag{50}$$

where Wo(g/m^3) means the quantity of the water droplets in the free stream.

Next, assuming that the water droplets starting at Lyo.1 and Lyo.2 reach the locations Lys.1 and Lys.2 on the obstacle, the total area per unit length of span can be written

$$A = L \cdot (Ys.2 - Ys.1) \tag{51}$$

Therefore, the mean rate of water droplet impaction per unit area becomes

$$\frac{G}{A} = \frac{\bar{V}o \cdot Wo \cdot L \cdot (Yo.2 - Yo.1)}{L \cdot (Ys.2 - Ys.1)} \tag{52}$$

Further, approaching Ys.1 to Ys.2, the limiting value of G/A of the water droplets becomes at the distance $y = Ys \cdot L$ as follows

$$\lim_{A \to o} \frac{G}{A} = \bar{V}o \cdot Wo \cdot \frac{dYo}{dYs} \tag{53}$$

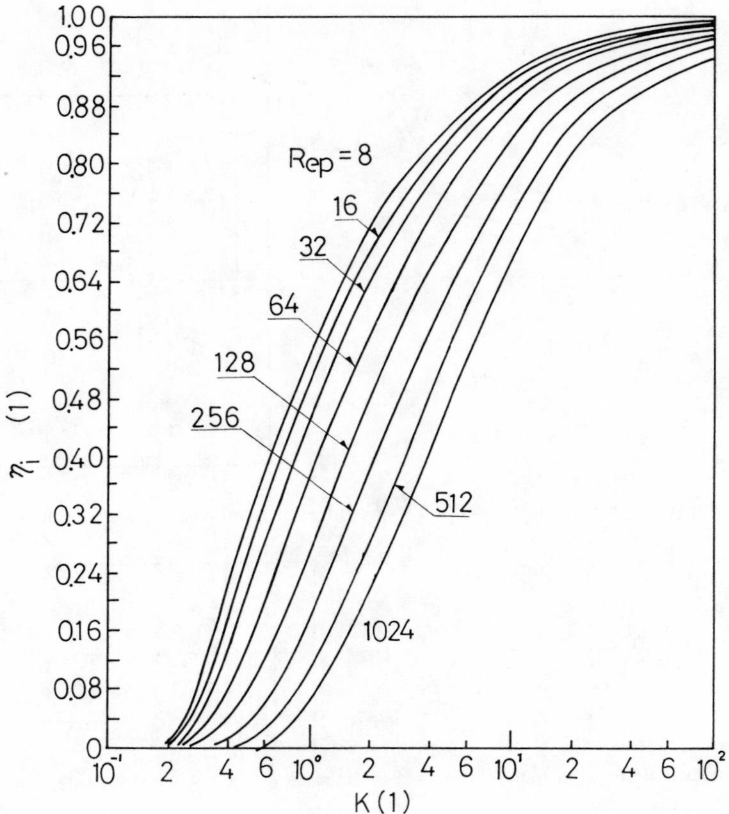

FIGURE 17. Theoretically calculated results of the impaction efficiency for the rectangular obstacle.

And then the fractional impaction efficiency η_{if} may be defined as

$$\eta_{if} = \frac{\lim (G/A)}{\bar{V}o \cdot Wo} = \frac{dYo}{dYs} \quad (54)$$

Here, the total impaction quantity W of the water droplets on the surface of the rectangular obstacle per unit length of span can be described as

$$W = 2\int_0^L \bar{V}o \cdot Wo \cdot \eta_{if} \cdot dy = 2 \cdot L \cdot \bar{V}o \cdot Wo \int_0^1 \frac{dYo}{dYs} \cdot dYs =$$

$$= 2 \cdot L \cdot \bar{V}o \cdot Wo \left\{ f(1) - f(0) \right\} \quad (55)$$

and then the value of f(0) corresponding to Yso = 0 becomes zero, so Equation 55 becomes

$$W = 2 \cdot L \cdot \bar{V}o \cdot Wo \cdot f(1) \quad (56)$$

Consequently the collection efficiency η_i can be defined as

$$\eta_i = \frac{W}{2 \cdot L \cdot \bar{V}o \cdot Wo} = f(1) \quad (57)$$

Figure 17 shows the theoretically calculated results of the impaction efficiency η_i for the rectangular obstacle. From this figure, we find that η_i strongly depends on the parameter K in comparison with Reynolds number Rep.

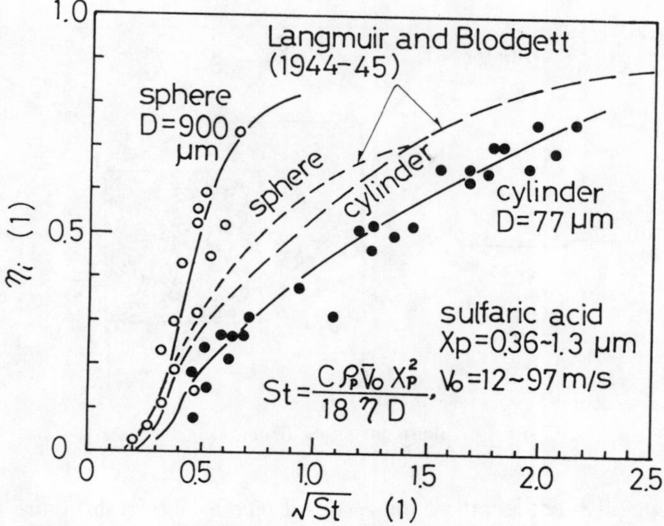

FIGURE 18. Experimental impaction efficiency of sulfaric acid aerosols.

C. Experimental Results of the Inertial Impaction

Experimental efficiency curves for impaction by inertia on cylindrical and spherical collectors are shown in Figure 18 by the experiment of Ranz and Wong (1952).[10] As shown in Figure 18, the diameter of a spherical collector is D = 900 μm and that of a cylinder is 77 μm. Then the diameter of aerosol is Xp = 0.36 to 1.3 μm and the relative velocities of aerosol stream are from Vo = 12 m/s to Vo = 97 m/s.

Those curves for impaction by inertia show a characteristic S-shape. As the ratio of particle diameter Xp to collector diameter D was never greater than Xp/D = 0.017, impaction by interception could be considered negligible. Impaction by an electrostatic mechanism was also improbable. Therefore, Figure 18 should represent impaction by the inertia mechanism alone.

The curve of the cylindrical collector closely follows the theoretical curve of Langmuir and Blodgett (1944—45) for low efficiency. Likewise, the curve of the spherical collector follows closely the theoretical curve of Langmuir and Blodgett for low efficiency.

IV. RESISTANCE OF FILTERS

A. The Kozeny–Carman Equation

The fluid flow in the fabric filter shows very complicated flow patterns which do not follow straight paths, but torturous paths in passing from layer to layer. Therefore, we assume that the gas flow in the representative converted pipe, named "hydraulic mean depth r_f", is defined as

$$r_f = \frac{\text{Af(sectional area of pipe)}}{\text{Lf(length of wet periphery)}} \tag{58}$$

If the pipe shape is circular of diameter Df, then the hydraulic mean depth r_f becomes Df/4 due to $Af = \pi \cdot D^2 f/4$ and $Lf = \pi \cdot Df$.

Here assuming that the thickness of the filter is H, then the hydraulic mean depth r_f becomes

$$r_f = \frac{\text{sectional area of a pipe} \times \text{Le}}{\text{length of wet periphery} \times \text{Le}} = \frac{\text{space volume of filter (Vs)}}{\text{surface area of filter (s)}} \tag{59}$$

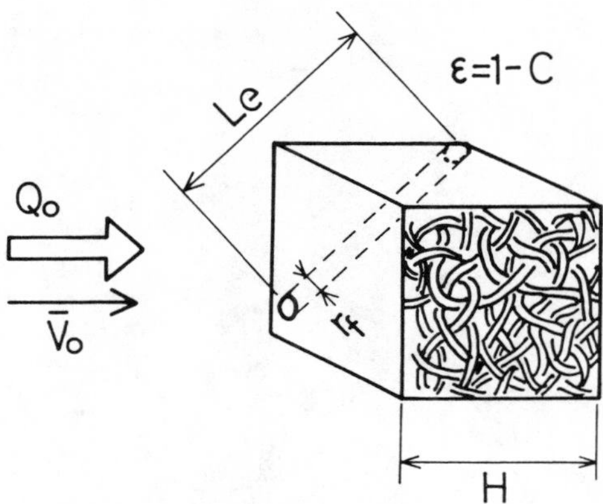

FIGURE 19. Illustration of the fluid flow in the fabric filter.

where Le is the equivalent length of the assumed pipe as shown in Figure 19.

The total volume Vt of the filter is composed of the volume Vf of the fabric filter and the space volume Vs, and also introduces the void ϵ defined as

$$\epsilon = \frac{Vs}{Vt} = \frac{Vt - Vf}{Vt} = 1 - \frac{Vf}{Vt} = 1 - C \tag{60}$$

the hydraulic mean depth r_f can be written

$$r_f = \frac{\epsilon \cdot Vt}{S} = \frac{\epsilon}{1-\epsilon} \cdot \frac{Vf}{S} = \frac{\epsilon}{1-\epsilon} \cdot \frac{1}{Sf'} = \frac{Df}{4} \tag{61}$$

where Sf′ means the surface area per unit volume of the fiber material. Here, assuming that the fluid flow in this pipe is a laminar flow, the pressure drop for the length H can be written by the Hagen-Poiseuille equation as

$$\Delta p = \frac{32 \cdot \eta \cdot \bar{V} \cdot Lf}{(4\, r_f)^2} = 2 \cdot \eta \cdot H \cdot \bar{V} \left(\frac{1-\epsilon}{\epsilon}\right)^2 \cdot Sf'^2 \tag{62}$$

Now the relationship between $\bar{V}o$ and $\bar{V}$ may be written

$$\bar{V} = \frac{\bar{V}o \cdot Le}{\epsilon \cdot H} \tag{63}$$

consequently Equation 62 can be transformed as

$$\Delta p = 2 \cdot \eta \cdot H \cdot \bar{V}o \cdot \frac{Le}{\epsilon \cdot H} \cdot \left(\frac{1-\epsilon}{\epsilon}\right)^2 \cdot Sf'^2 \tag{64}$$

Therefore, Equation 64 becomes

$$\Delta p = 2 \cdot \eta \cdot \bar{V}o \frac{C^2}{(1-C)^3} Sf'^2 \tag{65}$$

Further, Equation 65 can be transformed as

$$\Delta p = k \cdot \frac{4C^2}{(1-C)^3} \cdot \frac{\eta \cdot H \cdot Qo}{A \cdot R^2} \tag{66}$$

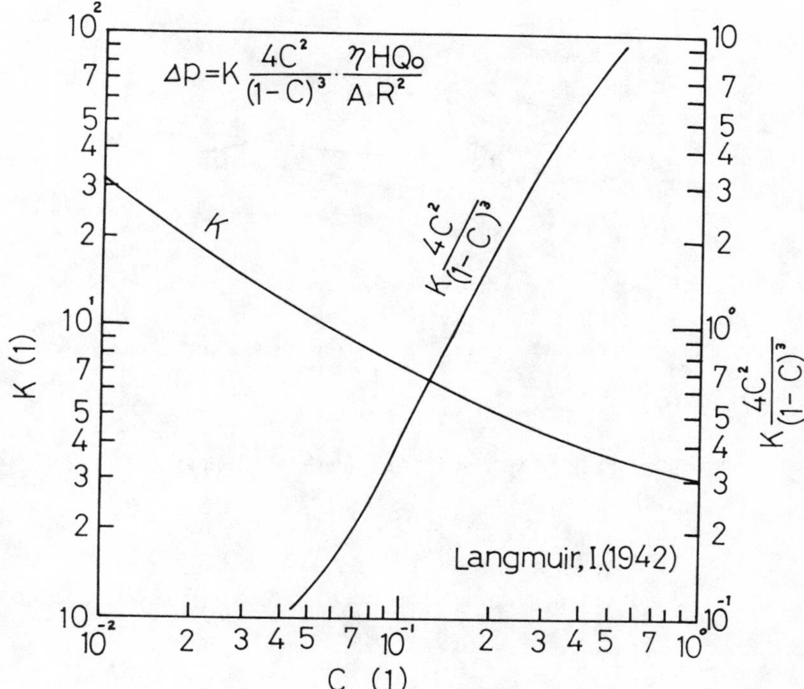

FIGURE 20. Relationship between Kozeny constant and packing density for fiber air filter.

where k is named the Kozeny constant and a relationship between Kozeny constant k and packing density C for fiber air filter by Langmuir[11] is shown in Figure 20.

From this figure, we can find that k does not vary much for $C \geqq 0.5$, but increases rapidly as the packing density falls below $C = 0.2$, which means that the resistance of loosely packed filters is much greater than would be indicated by the Kozeny-Carman equation.

On the other hand, Langmuir derived the equation of the pressure drop as

$$\frac{\Delta p \cdot A \cdot R^2}{\eta \cdot Qo \cdot A} = \frac{1.4 \times 4C}{-\ln C + 2C - \left(\frac{C^2}{2}\right) - \left(\frac{3}{2}\right)} \tag{67}$$

He considered that the resistance of the bed would be increased by a factor 1.4 if the cylinders lay across the flow direction.

Then the experimental expression of Δp which fits the data of Figure 20 by Davies, can be written

$$\frac{\Delta p \cdot A \cdot R^2}{\eta \cdot Qo \cdot H} = 1.6 \cdot C^{3/2}(1 + 56 \cdot C^3) = f_D(C) \tag{68}$$

B. Theoretical Calculation of Pressure Drop

The drag force per unit length of a cylinder transverse to the flow, as shown in Figure 21, is denoted by F as

$$F = C_D \cdot 2R \cdot \frac{1}{2}\rho \cdot \bar{V}o^2 \tag{69}$$

where C_D is the drag coefficient. The work done against the fluid resistance by a length l of cylinder per unit time can be written

$$F \cdot l \cdot \bar{V}o \tag{70}$$

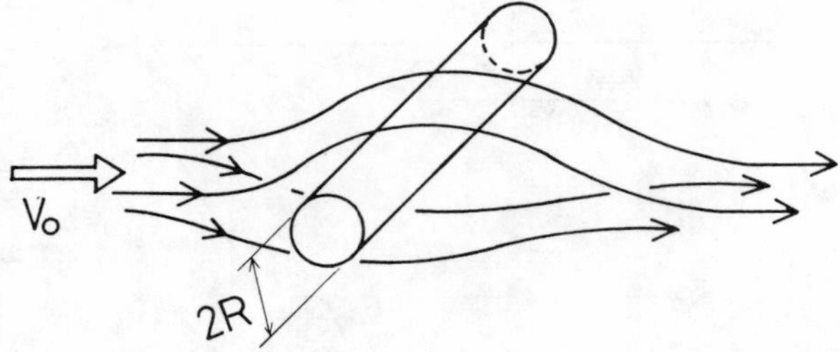

FIGURE 21. Illustration of fluid flow around a circular cylinder.

Then if the cylinder is distributed in a filter, this rate of working is also equal to

$$\Delta p \cdot Qo = \Delta p \cdot \bar{V}o \cdot A \tag{71}$$

Therefore, equating the two expressions of Equations 70 and 71, we can obtain

$$\Delta p = \frac{F \cdot l}{A} = \frac{F \cdot H \cdot C}{\pi \cdot R^2}$$

The force F′ on the unit length of an isolated fiber is given by Lamb's equation

$$F' = \frac{4 \cdot \pi \cdot \eta \cdot \bar{V}}{2.0022 - \ln \cdot Re} \tag{72}$$

provided that Re is very small.

In the case of random fiber arrangement, Happel (1959)[12] obtained a formula for flow along the cylinders

$$\frac{\Delta p \cdot A \cdot R^2}{\pi \cdot Qo \cdot H} = \frac{4C}{-\ln C + 2C - \left(\frac{C^2}{2}\right) - \frac{3}{2}} = f_{hp}(C) \tag{73}$$

and for flow transverse to the cylinder

$$\frac{\Delta p \cdot A \cdot R^2}{\pi \cdot Qo \cdot H} = \frac{8.C}{-\ln C - \left(\frac{1 - C^2}{1 + C^2}\right)} = f_{ht}(C) \tag{74}$$

It will be seen that Equation 73 for flow parallel to the axis of the cylinders is identical with the expression obtained by Langmuir, namely Equation 67 without factor 1.4.

Another way of expressing the boundary conditions and solving the Navier-Stokes equations for flow transverse to a set of parallel cylinders was used by Kuwabara (1959).[13] His general concept was the same as that of Happel, but the boundary condition no longer assumed absence of shearing stress at the outer cylindrical surface enclosing and coaxial with each fiber. From his theory we obtain Kuwabara's theory

$$\frac{\Delta p \cdot A \cdot R^2}{\pi \cdot Qo \cdot H} = \frac{8.C}{-\ln C + 2C - \left(\frac{C^2}{2}\right) - \frac{3}{2}} = f_k(C) \tag{75}$$

for flow transverse to a random assembly of parallel cylinders. The expressions of Equations 68, 73, 74, and 75 are shown in Figure 22.

The detailed mathematical description of the pressure drop of filters at small Knudsen number Kn = λ/R is written in the book *Aerosol Science,* edited by C. N. Davies[14] (see Chapter 9 by J. Pich, "Theory of Aerosol Filtration by Fibrous and Membrane Filters") as

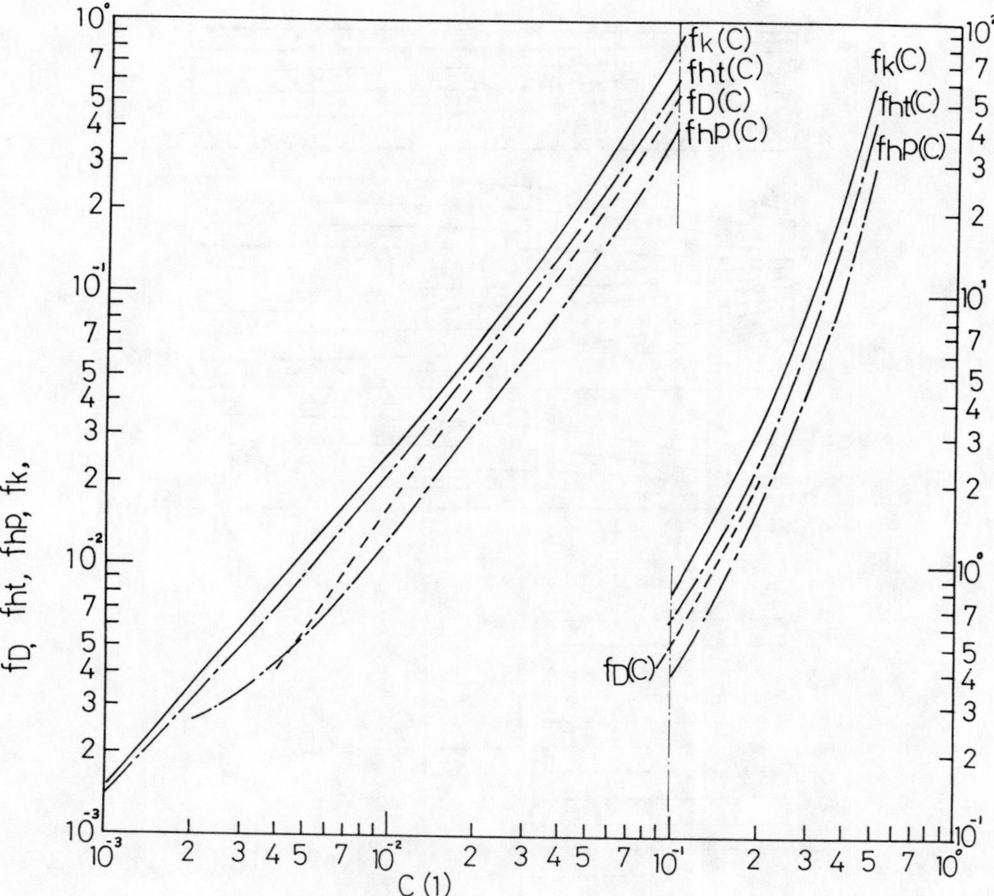

FIGURE 22. Numerical expressions of Equations 68, 73, 74, and 75.

a solution to the Navier-Stokes equations for the two-dimensional incompressible steady flow

$$\Delta\Delta\Psi = 0 \tag{76}$$

where Ψ is a stream function and can be written

$$\Delta = \frac{\partial^2}{\partial r^2} + \frac{1}{r}\cdot\frac{\partial}{\partial r} + \frac{1}{r^2}\frac{\partial^2}{\partial \theta^2}$$

And then a particular solution of Equation 76 is supposed as

$$\Psi = f(r)\cdot\sin\theta \tag{77}$$

Kovasznay (1948)[5] did a numerical analysis of the stream lines for laminar flow behind a two-dimensional grid. His paper presented an exact two-dimensional solution of the Navier-Stokes equations with a periodicity in one direction which may be represented by the wake of a two-dimensional grid. Figure 23 (a) for $\overline{V}o\cdot M/\upsilon = 40$ represents a flow past a grid choosing the stagnation points (denoted by S in those figures) at X = 0. A pair of bound eddies occur behind the single elements of the grid. The dotted stream lines correspond to the half-value of the stream function between two full lines. Figure 23 (b) shows the other

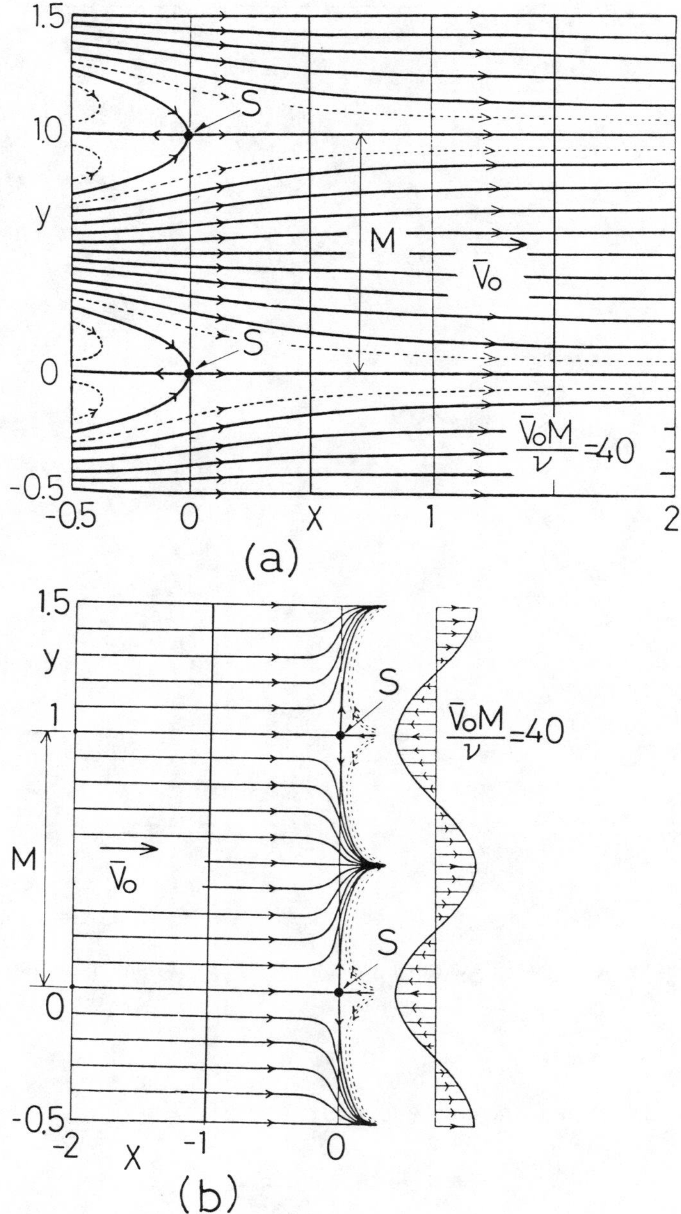

FIGURE 23. Illustration of the flow patterns past a grid.

solution of the stream lines by the vorticity equation for $\overline{V}o \cdot M/\upsilon = 40$. The rate of change of the flow is very great.

This figure may represent a flow of alternative vortices superposed on a main flow perpendicular to their plane. Those figures may describe the motion of fluid behind a grid consisting of equally spaced parallel rods or strips.

C. Kuwabara Flow Model About Cylinders

In order to simulate the condition of random and homogeneous distribution of bodies, Kuwabara (1959)[13] took a model such that, in the coordinate system moving with the mean

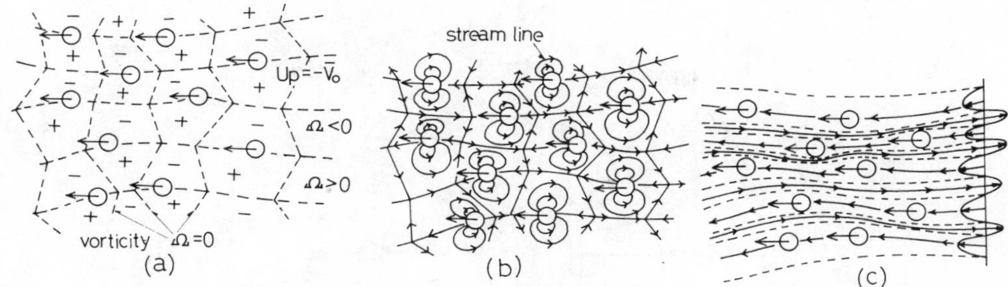

FIGURE 24. Detailed illustration of the Kuwabara-flow model.

flow velocity $\overline{V}o$, each one of the bodies is moving with the negative velocity $-\overline{V}o$ in a certain closed surface on which both the normal fluid velocity and vorticity Ω vanish.

In the case of the parallel circular cylinders, he assumed that

1. All the circular cylinders have the same radius R and the same velocity $\overline{U}p = -\overline{V}o$.
2. They are parallel to each other and perpendicular to the mean flow.
3. They are distributed at random and homogeneously. Further, the mean flow is directed from left to right. On the upper side of a cylinder the vorticity Ω would be negative and on the lower side positive.

Therefore midway between the neighboring cylinders, there will be boundary lines of zero vorticity ($\Omega = 0$) as shown by the dotted lines in Figure 24 (a).

Then Kuwabara assumed that the stream lines are closed in the neighborhood of the cylinder in the reference system which is moving at the mean velocity, as shown in Figure 24 (b). On the other hand, the flow pattern of a series of infinitely long strips consisted of fluid with alternating flow directions located side by side, as shown in Figure 24 (c). However, he said that such a flow would be improbable.

From the above stated assumptions, Kuwabara took the following mathematical model which consisted of an imaginary circular cylinder of radius b enclosing and coaxial with the given circular cylinder of radius R. On the imaginary circular cylinder, the vorticity Ω and the normal component of velocity were assumed to be zero, while on the surface of the solid circular cylinder, the velocity should be equal to minus the mean flow velocity Vo as shown in Figure 25. Then he took the cross-sectional area $\pi \cdot b^2$ of the imaginary circular cylinder to be equal to the "free" area corresponding to each solid cylinder, namely,

$$\pi \cdot b^2 = \frac{1}{n} \tag{78}$$

where n was the number of the solid cylinders per unit area. The fundamental equation for the two-dimensional steady flow motion to Stokes approximation was written

$$\Delta\Delta\Psi = 0 \tag{79}$$

where Ψ was the stream function of cylindrical coordinates (r,θ). The velocity $(Vr, V\theta)$ and the vorticity Ω were given as

$$Vr = \frac{1}{r} \cdot \frac{\partial \Psi}{\partial \theta} \qquad V\theta = -\frac{\partial \Psi}{\partial r} \tag{80}$$

$$\Omega = \frac{\partial V\theta}{\partial r} + \frac{V\theta}{r} - \frac{1}{r}\frac{\partial Vr}{\partial \theta} = -\Delta\Psi \tag{81}$$

where

$$\Delta = \frac{\partial^2}{\partial r^2} + \frac{1}{r} \cdot \frac{\partial}{\partial r} + \frac{1}{r^2}\frac{\partial^2}{\partial \theta^2} \tag{82}$$

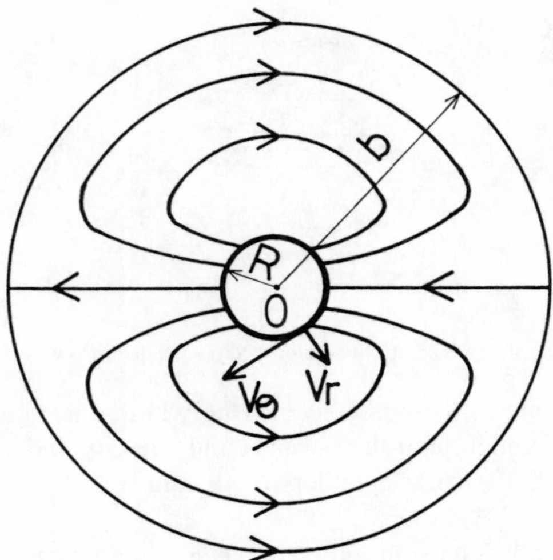

FIGURE 25. Kuwabara-flow model.

The boundary conditions of his model became:

$$Vr = V\theta = 0 \qquad \text{for} \qquad r = 1 \tag{83}$$

$$Vr = U\cdot\cos\theta \qquad \text{for} \qquad r = b/R \tag{84}$$

$$\Omega = 0 \qquad \text{for} \qquad r = b/R \tag{85}$$

in the reference system in which the cylinders were at rest, where $\overline{V}o$ was the velocity of the mean flow parallel to the original line $\theta = 0$, and r had been made dimensionless by taking it as an unit. The particular solution of Equation 79 could be written

$$\Psi = \left(A\cdot\frac{1}{r} + B\cdot r + C\cdot r\cdot \ln r + D\cdot r^2\right)\cdot \sin\theta \tag{86}$$

where constants A, B, C, and D could be determined by the boundary conditions above stated as

$$A = \frac{1}{2}\cdot\frac{\overline{V}o}{K}\left\{1 - \frac{1}{2}\left(\frac{b}{R}\right)^{-2}\right\} \tag{87}$$

$$B = -\frac{1}{2}\cdot\frac{\overline{V}o}{K}\left\{1 - \left(\frac{b}{R}\right)^{-2}\right\} \tag{88}$$

$$C = \frac{\overline{V}o}{K} \tag{89}$$

$$D = -\frac{1}{4}\,\overline{V}o\Big/\left\{K\left(\frac{b}{R}\right)^2\right\} \tag{90}$$

$$K = \ln\left(\frac{b}{R}\right) - \frac{3}{4} + \left(\frac{b}{R}\right)^{-2} - \frac{1}{4}\cdot\left(\frac{b}{R}\right)^{-4} \tag{91}$$

The drag force X per unit length of a circular cylinder could be calculated by Imai's formula which was an extension of Blasius formula to the case of a viscous flow. In this case, the formula could be written

$$X = -2\eta\int_c \frac{\partial\Omega}{\partial\overline{Z}}\cdot\overline{Z}\cdot d\overline{Z} \tag{92}$$

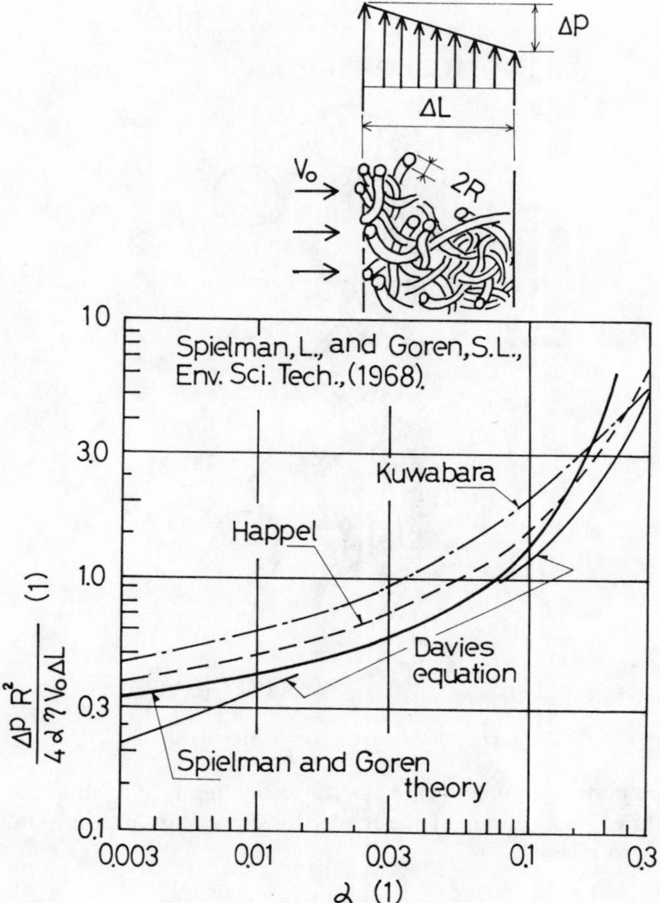

FIGURE 26. Dimensionless pressure gradient and volume fraction fibers for fiber axes all perpendicular to direction of flow comparing cell models and present theory with Davies empirical relation.

where $\overline{Z} = r \cdot \exp(-i\theta)$ and C denoted the contour of the cross-section of the solid cylinder. Thus, he got the equation

$$X = 4 \cdot \pi \cdot \eta \cdot \frac{\overline{V}o}{\left\{ \ln\left(\frac{b}{R}\right) - \frac{3}{4} + \left(\frac{b}{R}\right)^{-2} - \frac{1}{4}\left(\frac{b}{R}\right)^{-4} \right\}} \tag{93}$$

And also the drag coefficient C_D of the cylinder could be obtained as

$$C_D = \frac{X}{\frac{1}{2} \cdot \rho \cdot \overline{V}o^2 \cdot (2R)} \tag{94}$$

Then when the value of (b/R) became infinity, X became zero which was a paradox of Stokes.

Further, Spielman and Goren (1968)[15] applied Briukman's model for flow-through unconsolidated porous media to flow-through fibrous media and predicted pressure drop for several different internal arrangements of fibers, as shown in Figure 26.

The essence of the model is that, on the average, the fluid in proximity to an obstacle embedded in a porous medium experiences, in addition to the usual force terms, a body

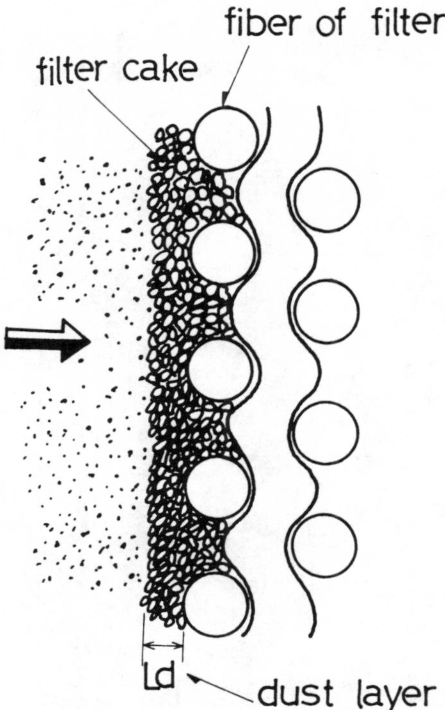

FIGURE 27. Dust load with lapse of time on the filter.

damping force proportional to the local velocity accounting for the influence of neighboring objects on the flow. Implications of the model for correlating pressure drop and filtration efficiency were also given.

V. EFFICIENCY OF FABRIC FILTER

A. Pressure Drop

The pressure drop Δp of the filter is in general expressed by the sum of the pressure drop Δp_o of the fabric filter and also of the pressure drop Δp_d of the dust layer on the filter as

$$\Delta p = \Delta p_0 + \Delta p_d = \zeta \cdot \eta \cdot \bar{v} = (\zeta_0 + m \cdot \alpha)\eta \cdot u \tag{95}$$

where Δp_o can be represented as

$$\Delta p_0 = \zeta_0 \cdot \eta \cdot u \tag{96}$$

and also Δp_d can be represented as

$$\Delta p_d = \zeta_d \cdot \eta \cdot u \tag{97}$$

Here ζ_o (1/m) represents the resistance coefficient of the filter, ζ_d (1/m) represents the resistance coefficient of the dust layer, m represents the dust load (kg/m^2) on the filter, and α represents the specific resistance (m/kg)of the dust layer.

As shown in Figure 27, increasing the dust load with lapse of time on the filter decreases the filtering velocity of gas. From the practical point of view, the pressure drop Δp_d in the dust layer on the filter was represented by Billings (1970)[16] as

Table 2
NUMERICAL FACTORS OF FILTERS (BY BILLINGS)

Material factor	Crushed material	Fume	Fly-ash	Irregular form	Soft and transformed material				
Ks (1)	10	0.05	4	3	0.2				
Factor of filter	**Filament**	**Staple spun**	**Felt**						
Kf (1)	1	0.5	0.25						
Permeability	**5**	**10**	**15**	**20**	**25**	**30**	**35**	**40**	**45**
Kp (cm/s)	1.3	1.2	1.1	1.0	0.9	0.8	0.7	0.6	0.5

$$\Delta p_d = 10 \cdot K \cdot Ci \cdot \overline{V}^2 \cdot t, \text{ (Pa)} \tag{98}$$

$$K = 16\, Ks \cdot Kf \cdot Kp\left(\frac{\overline{V}}{Xp^2}\right) \tag{99}$$

where Ci (g/m^3) represents the dust concentration at the inlet section, $\overline{V}$ (m/s) represents the filtering velocity, t (min) is lapse of time of filtering, Ks (1) represents the material factor of dust, Kf (1) represents the factor of filters, Kp (cm/s) represents the permeability of the filter and Xp (μm) represents particle diameter, as shown in Table 2.

Here we try to estimate the pressure drop Δp_d using Billings' equation. Using a staple spun filter of permeability 15 cm/s and, if the inlet dust concentration of fly ash of diameter $\overline{Xp} = 10\ \mu$m is Ci $= 5$ g/m^3 and the filtering velocity is $\overline{V} = 1.5$ m/min, we calculate the pressure drop Δp_d after 30 min of filtering time.

First of all we must calculate the value of K as

$$K = \frac{16}{(10)^2} \times 4 \times 0.5 \times 1.1 \times 1.5 = 0.528$$

therefore the pressure drop Δp_d becomes

$$\Delta p_d = 10 \times 0.528 \times 5 \times (1.5)^2 \times 30 = 1782 \text{ Pa}$$

In addition, we must consider the effects of the gas humidity concerning the pressure drop. In general, increasing the adhesive forces between the particles with increasing the gas humidity, the dust layer on the filter becomes a large void dust layer. Therefore, the pressure drop Δp must be decreased in comparison with that of the dry dust-laden gas flow. However, this consideration is not always true.

Figure 28 shows the experimental results (by Mori) of the pressure drop Δp related to the relative humidity of gas.[17] Experimental conditions are as follows: using a felt filter, the mean diameter Xp of loam is 8 μm, the filtering velocity $\overline{V}$ is 2.7 m/min. From this figure, we can find that the pressure drop Δp does not depend on the relative humidity within those humidities. Consequently, we must deeply consider the mechanical property of the deposition of dust on the filter.

Figure 29 shows the relationship between Δp and V which has a parameter dust load of m (g/m^2), as in Mori's experiments. He used a felt filter and $CaCO_3$ dust. From his experimental results, Δp is nearly proportional to the filtering velocity $\overline{V}$ (m/min) and Δp can be nearly expressed as

$$\Delta p\,(\text{Pa}) = 72 \cdot \overline{V}^{-0.8} \cdot m^{0.4\sqrt{\overline{V}}} \tag{100}$$

for the conditions $1.5 \leqq \overline{V} \leqq 3$ m/min and $40 \leqq m \leqq 300$ g/m^2.

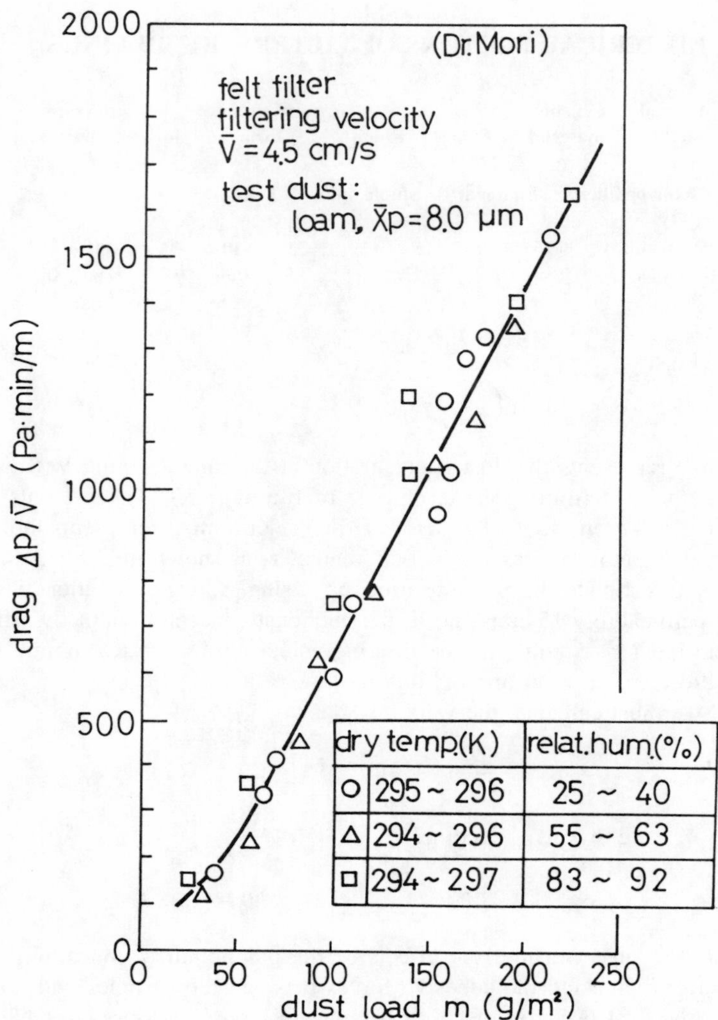

FIGURE 28. Experimental results of the pressure drop related to the relative humidity of gas.

B. Collection Efficiency

The bag filter system has one of the highest degrees of collection efficiency. But with improper selection of the filter material, or an inadequate driving condition, the collection efficiency will be decreased. On the contrary, taking notice of the above, we can keep a constant high collection efficiency so that the dust concentration passing through the bag filter is suppressed below 10 to 30 mg/m^3. At the same time, due to increasing the adhesive forces of fine solid particles, the collection efficiency does not strongly depend on the particle size.

In order to recognize the adhesive process of dust in the filtration process, Eliseev (1957)[18] investigated the deposition of fine particles on a single cylindrical wire placed in a flow of dust-laden gas. The formation of the dust deposition on the individual cylindrical wire for flow velocity Vo = 1 m/s is shown in Figure 30.

The precisely visible local side growths of lead and zinc oxide particles about Xp ≒ 1

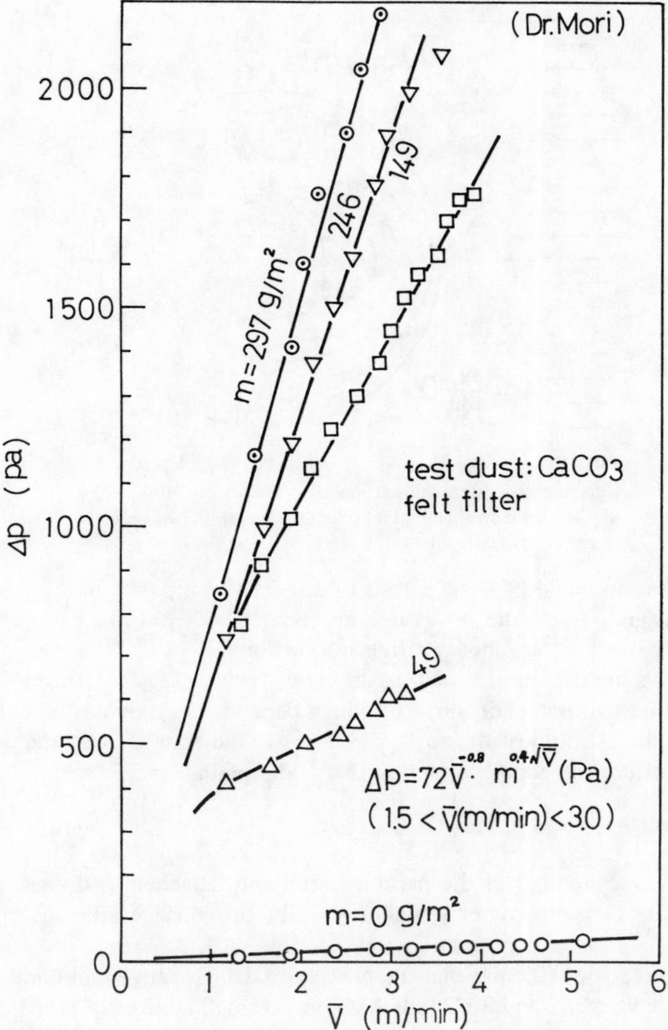

FIGURE 29. Relationship between the pressure drop and the filtering velocity.

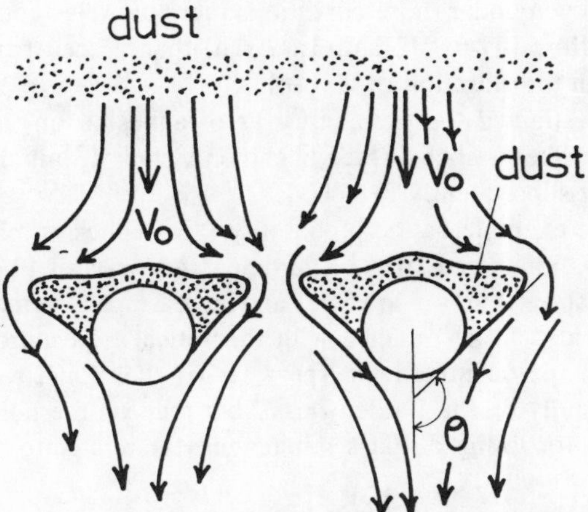

FIGURE 30. Formation of an adhesive dust layer on the individual wires by Eliseev.

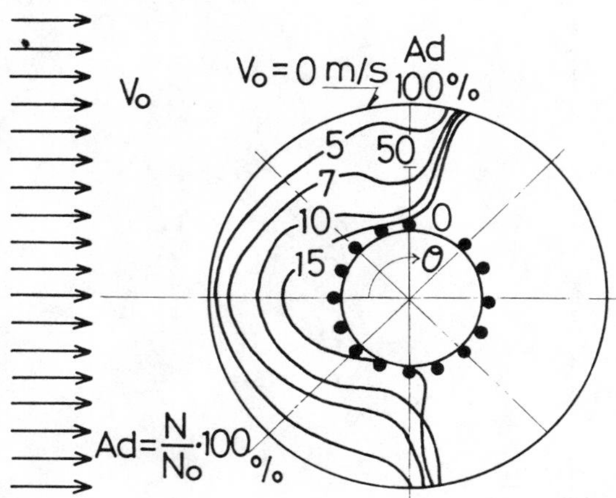

FIGURE 31. Adhesion number Ad associated with the detachment of loss particles of diameter Xp = 40 to 100 μm by air flow from a cylindrical porcelain surface arranged vertically in an aerodynamic tube.

μm are directed at an angle of $\theta \doteqdot \pm 1.92$ to 2.09 rad to the axis of the gas flow. When more dust-laden gas passes, the growths may come together forming a continuous layer which plays the part of a "secondary" filtering medium.

Figure 31 shows the detachment of the particles (Xp = 40 to 100 μm) on the surface of the cylinder (cylindrical porcelain surface) which depends not only on the pure air velocity Vo, but also on the position of the surface relative to the flow axis by the experiments of Tekenev (1962). Here the adhesive number Ad is defined as

$$\mathrm{Ad} = \frac{\mathrm{N}}{\mathrm{No}} \times 100 \quad (\%) \tag{101}$$

where No means the number of the particles originally attached to the test surface and N means the number of the particles remaining on the test surface after the application of a given force.

From this figure, you will find that the maximum detachment (minimum Ad) occurs for the value of angle $\theta = \pi/2$ and $3\pi/2$ rad. Also only a small number of particles are detached from the front ($\theta = 0$ rad) and none at all from the back at the velocities Vo (= 0 to 15 m/s) in this experiment. Then with increasing flow velocity, the adhesion number Ad diminishes. However, even under those conditions more particles come away from surfaces placed parallel to the flow ($\theta = \pi/2$ and $3\pi/2$ rad) than placed frontally. This fact is of particular importance in the filtration of aerosols.

Löffler (1977)[19] investigated the probability Pr of adhesion on the surface of the fiber filter. When a particle strikes a fiber surface, it can only stick if, initially, it does not rebound and, secondly, it is not subsequently detached.

Figure 32 shows one example of the probability Pr of adhesion plotted against the fluid velocity Vo for various kinds and sizes of particles and types of fibers.

Rebounding starts at about Vo = 5 to 15 m/s and increasing the velocity Vo, the probability of adhesion drops quickly. In accordance with theoretical expectations, the probability of adhesion is lesser for the particle diameter Xp = 10 μm than for Xp = 5 μm. The particles also rebound more strongly on the Larder glass fiber than on the polyamide fiber.

The lower adhesion probability of glass spheres may be related to an effect of the contact geometry.

Irregular quartz particles may have multi-point or area contacts while only single-point contacts are possible with very smooth glass spheres.

Figure 33 shows the experimental results of the collection efficiency of a felt filter which

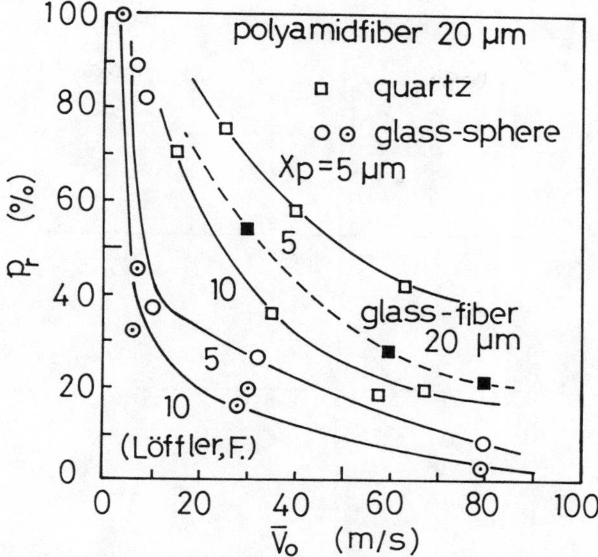

FIGURE 32. Probability Pr of adhesion for quartz particles and glass sphere particles on a 20 μm polyamide fiber.

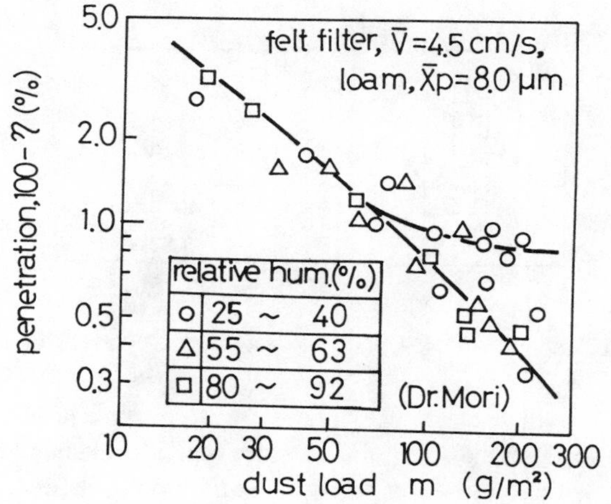

FIGURE 33. Relationship between the penetration and the dust-load.

is related to the dust load m (g/m^2) of loam $\overline{X}p$ = 8.0 μm. From this figure, the reader will recognize that the collection efficiency η does not depend on the relative humidity. The collection mechanism of the bag filter is mainly related to the filtration by the dust-layer which is deposited on the filter surface or is penetrated into the filter.

Early in filtration, diffusion and inertia depositions are strongly affected, but in the lapse of time an interception effect may be strongly related. Therefore, after dust dislodging, the dust layer on the filter becomes thinner due to the separation of the dust layer from the surface of the filter, and then the collection efficiency falls down temporarily. However, after beginning the filtration again, the collection efficiency will recover from the bad condition within several minutes.

C. Pressure Loss Control

There are three methods of controlling pressure loss across the bag filter, i.e., timer type, one-point pressure type, and two-point pressure type.[20] The timer type is the most popular

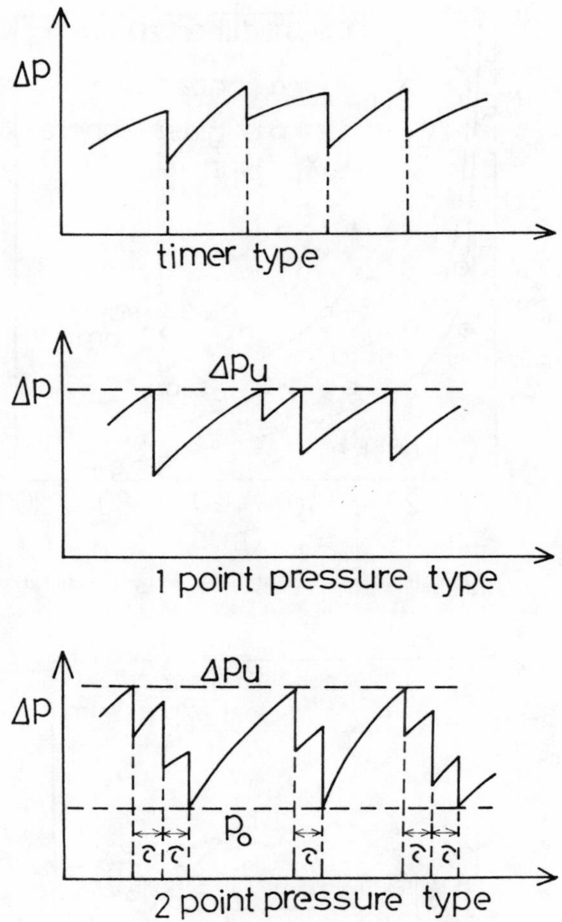

FIGURE 34. Three types of pressure drop controls of bag filters.

in general industry, in which cleaning operates for a constant period in the compartment with the highest dust load. In the one-point pressure type, the cleaning operation starts when the pressure loss across the fabric filter reaches a specified upper limit Δp_u. Also in the two-point pressure type, cleaning starts when the pressure loss reaches a fixed upper limit and stops when the loss falls to a fixed lower limit Δp_l.

The typical three patterns of pressure loss over a cycle for each of the control methods are shown in Figure 34.

Generally speaking, when the dust layer load on the bag filter reaches 0.1 to 0.3 kg/m^2, the pressure drop across the filter becomes $\Delta p = 1000$ to 2000 Pa. Therefore, if we do not carry out the treatment of the dust dislodging, we cannot continuously use the bag filter. The difficulty of dust dislodging depends on the combination of the filter materials and the kinds of dust, but the fabric filter is an easier treatment than a felt filter, and also the coarse particles is an easier treatment than fine particles.

On the other hand, if dust dislodging from the bag filter is done perfectly, the collection efficiency decreases in the same way as does the pure clean filter. Therefore, it is desirable to do the degree of the dust dislodging which may remain as a thin dust layer (primary dust layer) on the surface of the filter. As a quantitative expression of the pressure drop concerning the dust dislodging, we may define a residue fraction Λp as the ratio of the pressure drops

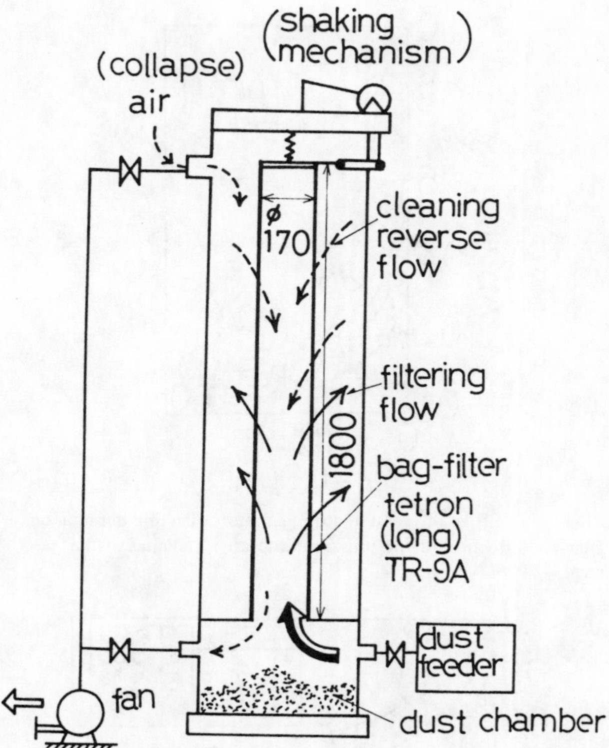

FIGURE 35. Experimental apparatus for bag filter systems.

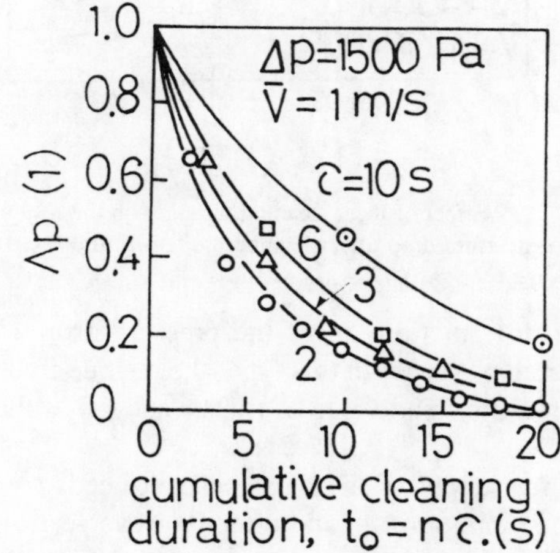

FIGURE 36. Effect of local cleaning duration on residue fraction of pressure drop after cleaning of mechanical shaking type.

after and before the dust dislodging. Iinoya and Makino (1977) have done experiments concerning the residue fraction of the pressure drop with the experimental apparatus of collapse and mechanical shaking types of bag filters, as shown in Figure 35.

Figures 36 and 37 show the residue fraction Λp of the pressure drop after cleaning the tetron (long) TR-9A filter by the mechanical shaking. From Figure 36, we can see that even if the long time of the cleaning duration is continued, the effect becomes scanty. Also from Figure 37, the residue fraction Λp decreases with increasing Δp.

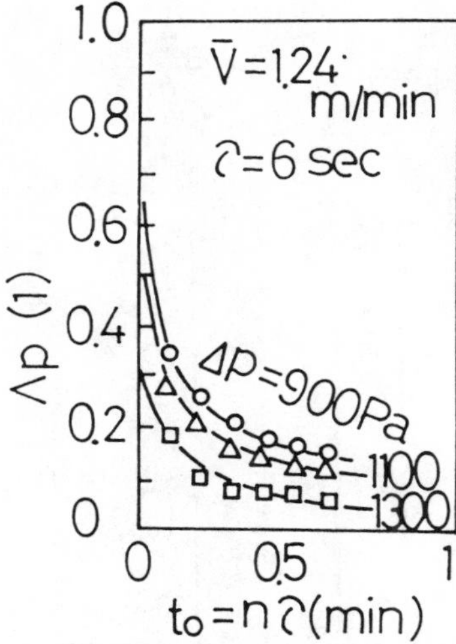

FIGURE 37. Effect of pressure drop just after collecting duration on the residence fraction of the pressure drop after cleaning of the mechanical shaking type.

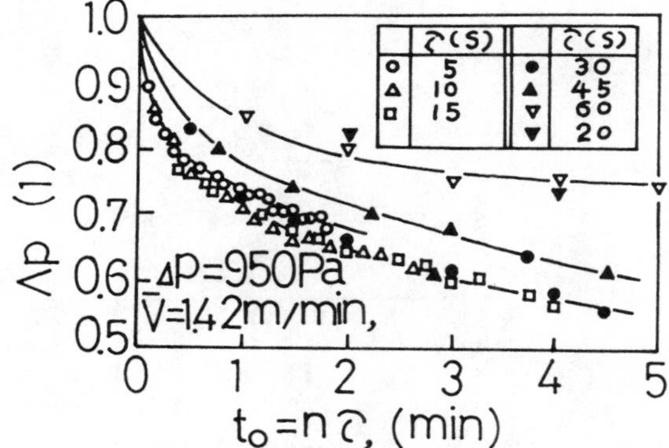

FIGURE 38. Effect of local cleaning duration on the residue fraction of pressure drop after cleaning the tetron filter by the collapse type ($\overline{V}r$ = 0.73, 1.07 m/min).

Figure 38 shows the residue fraction Λp of the pressure drop after cleaning the tetron (long) TR-9A filter by the collapse type. In this type, the residue fraction Λp has a minimum value at τ = 10 to 30 sec for any fixed value of t_o. The actions of the repeated reverse flow are effective.

Because of the large compartment of the bag house, it needs some waiting time before reaching the effective reverse pressure, or the effective reverse flow on the filter surface. Consequently, the optimal reversal time must be estimated from the volume of the compartment and from the value of the resistance or pipe connection.

D. Methods of Dust Dislodging

In general, there are four types of the dust dislodging systems, i.e., shaking, reverse air flow or collapse, reverse pulse-jet, and reverse jet. For a long time, the shaking and reverse flow (collapse) systems were used for cleaning fabric filters. Depending on the filter material, the shaking method is effective in comparison with the reverse flow method. But the lifetime of the filter material is short in comparison with the reverse flow system. Recently the pulsating reverse air and the pulse-jet methods have been applied.

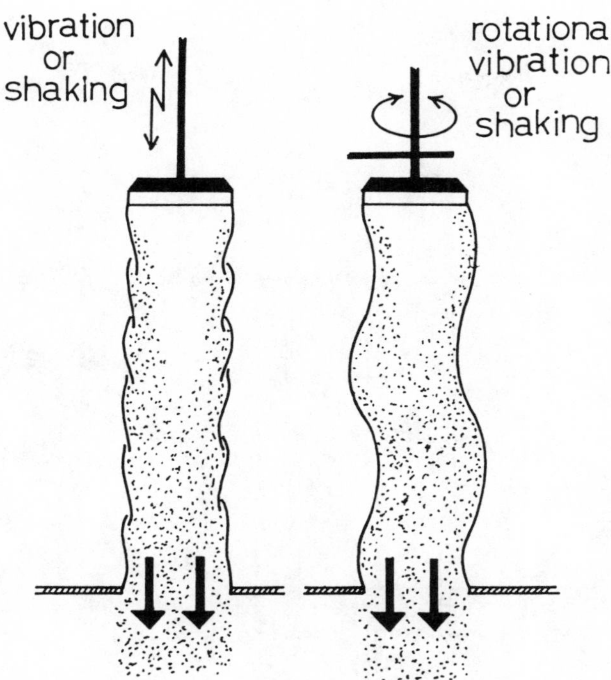

FIGURE 39. Mechanical cleaning method.

Figure 39 shows the shaking methods[21] where the fabric filter is vibrated for mechanical cleaning systems. The frequency of the vibration is several hundred revolutions per minute and the amplitude of the vibration is within 10 to 20 cm.

Figure 40 (a) and (b) show the reverse flow and the reverse pulse-jet cleaning systems. When the reverse flow is passed coercively through the filter material after stopping the filtering flow system, the collected dust layer on the surface of the filter becomes loose and the surface of the filter is deformed by the reverse gas flow. The dust layer on the filter receives a large strain and the dust layer lump is separated from the filter's surface.

The reverse pulse-jet cleaning system is a type which coercively flows the pulsating gas on the bag filters by opening and closing the values continuously.

This system of dust dislodging is very easy, due to the strong increasing and decreasing acceleration. Figure 41 shows the reverse jet system that instantaneously ejects the compressed air from a nozzle or slot on the surface of the cylindrical bag filter. Dust dislodging can be finished in a moment against the normal gas flow. This system can be designed from two to three times the ordinary filtering velocity and therefore becomes a compact type.

It is obvious from the results that electrostatic augmentation of fabric filters yields improvement in collection efficiency, but very little information concerning the effects of electrostatics on filter resistance is available. For example, it will be shown that by simply charging particles, the filtration rate per unit fabric area of a given fabric filter can be increased by a factor of 4. This implies that with the aid of electrostatics it might reasonably be expected that fabric filters can be reduced in size by a factor of 4, while achieving better collection efficiency.[22]

Figure 42 shows a practical example of the sketch and of the operation of an electrostatically aided fabric filter (APITRON) by Helfritch. Figure 43 shows the influence of the electrical energization on the pressure drop by Helfritch. When no high voltage is applied, the fabric

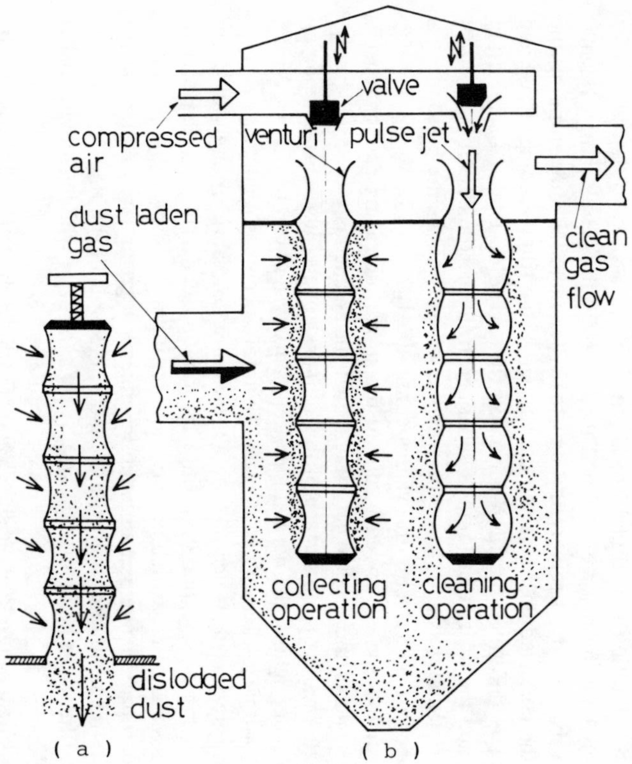

FIGURE 40. Cleaning systems of bag filters. (a) Reverse flow cleaning system; (b) reverse pulse-jet cleaning system.

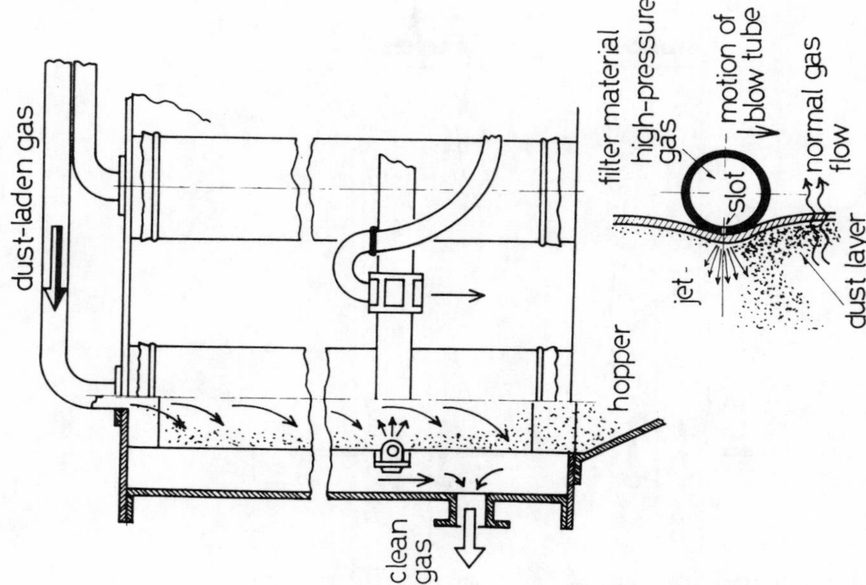

FIGURE 41. Reverse jet system of bag filters.

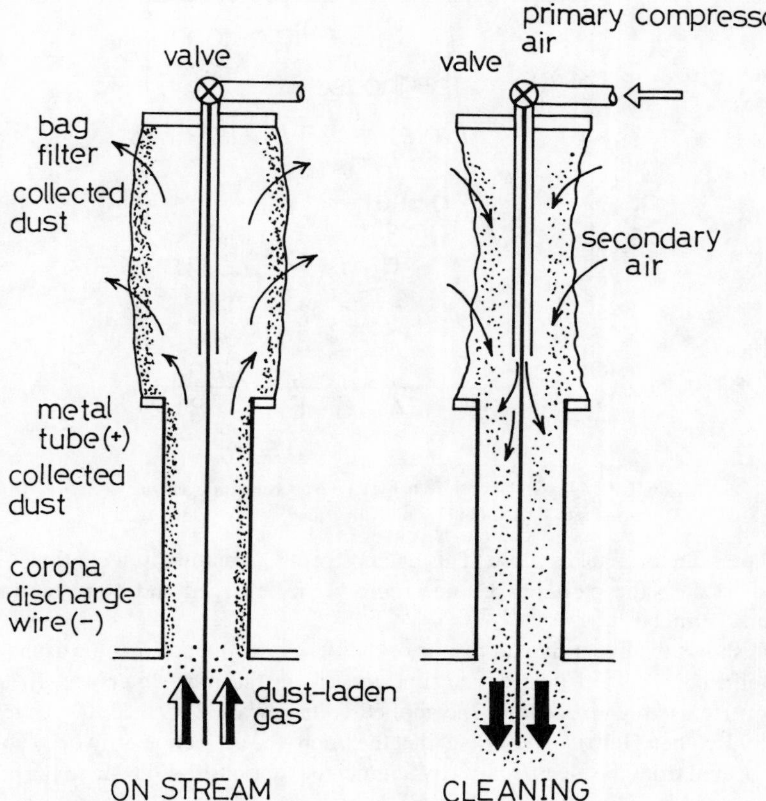

FIGURE 42. Electrostatic fabric filter.

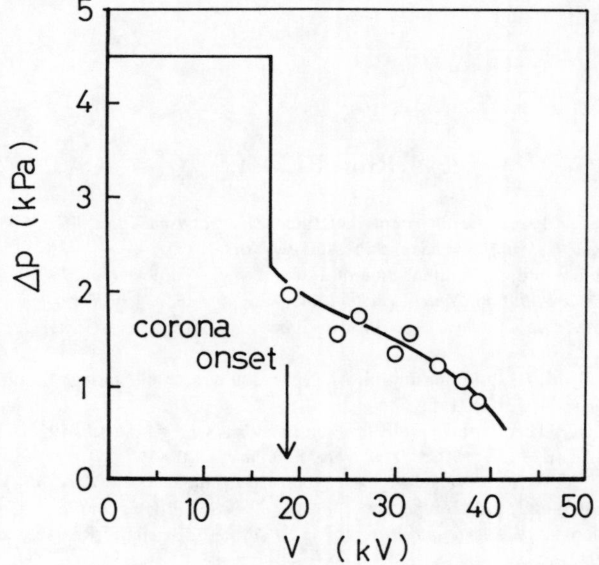

FIGURE 43. Influence of electrical energization on pressure drop.

filter behaves as a conventional continuous cleaning pulse-type bag house. When a corona discharge is generated and, at the same time the solid particles are charged, a sudden change

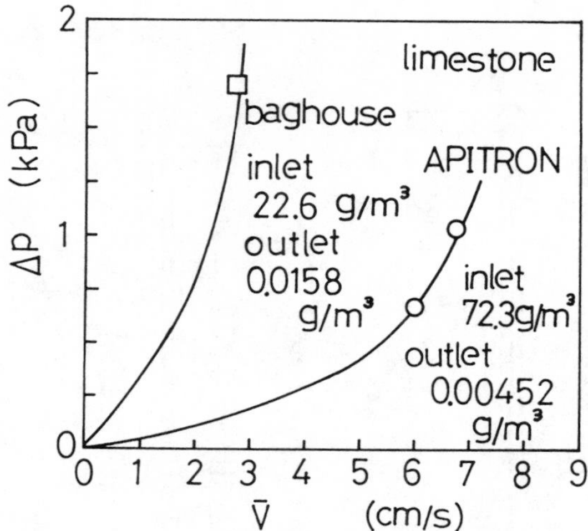

FIGURE 44. Relationships between the pressure drop and the filtration rate for the collection of limestone.

in filtration pressure drop takes place. This decrease of the pressure drop continues as voltage is increased because the particles become more highly charged and more particulates are deposited on the metal tube.

Figure 44 shows the experimental results of the field test conducted at the asphalt batching plant by Helfritch. The test dust of the limestone from the rock dryer was 10% less than Xp = 4 μm. He used Nomex® felt. The applied voltage was 32 kV and the gas temperature was T = 339 K. Then Helfritch described that the improved collection efficiency (APITRON) could be expected from the electrostatically augmented fabric filtration due to the polarization effects between charged particles and fibers.

REFERENCES

1. **Kaufmann, A.,** Die Fasestoffe für Atemschutzfilter, *Z. Ver. Dtsch. Ing.*, 8020, 593, 1936.
2. **Davies, C. N.,** *Air Filtration,* Academic Press, New York, 1933.
3. **Ranz, W. E. and Wong, J. B.,** Impaction of dust and smoke particles, *Ind. Eng. Chem.*, 44, 1371, 1952.
4. **Bethea, R. M.,** *Air Pollution Control Technology,* Van Nostrand, New York, 1978.
5. **Kovasznay, L. S.,** Laminar flow behind a two-dimensional grid, *Proc. Cambridge Philos. Soc.*, 44, 58, 1948.
6. **Yoshikawa, N., Emi, H., and Yasunami, M.,** Filtration of aerosols through fibrous packed bed, *Kagaku Kogaku,* (in Japanese), 33 (4), 1969.
7. **Friedlander, S. K.,** Theory of aerosol filtration, *Ind. Eng. Chemis.*, 1161, 1958.
8. **Lewis, W. and Brun, P. J.,** *NACA Tech. Note,* February, 3658, 1956.
9. **Milne-Thomson, L. M.,** *Theoretical Hydrodynamics,* Macmillan, New York, 1938.
10. **Ranz, W. E. and Wong, J. B.,** Impaction of dust and smoke particles, *Ind. Eng. Chem.*, 44, 1371, 1952.
11. **Langmuir, I.,** *Report on Smoke and Filters,* I.U.S. Office of Scientific Research and Development, No. 865, Part IV, 1942.
12. **Happel, J.,** Viscous flow relative to arrays of cylinders, *Am. Inst. Chem. Eng. J.*, 5, 174, 1959.
13. **Kuwabara, S.,** The forces experienced by randomly distributed parallel circular cylinders or spheres in viscous flow at small Reynolds number, *J. Phys. Soc. Jpn.*, 14, 527, 1959.
14. **Davies, C. N.,** *Aerosol Science,* Academic Press, New York, 1966.

15. **Spielman, L. and Goren, S. L.,** Model for predicting pressure drop and filtration efficiency in fibrous medium, *Environ. Sci, Tech.,* 2, 279, 1968.
16. **Billings, C. E.,** *Handbook of Fabric Filter Technology,* Vol. 1, U.S. Department of Commerce, Washington, D.C., 1970.
17. **Iinoya, K.,** *Dust Collection Engineering* (in Japanese), Nikankogyo-Shinbun Sha, Tokyo, 1980.
18. **Zimon, A. D.,** *Adhesion of Dust and Powder,* translated from the Russian by M. Corn. Plenum Press, New York, 1969.
19. **Löffler, F.,** Collection of particles in fibre filters, Proc. 4th Int. Clean Air Congr., Tokyo, May, 1977.
20. **Iinoya, K. and Makino, K.,** Pressure loss control of bag filters, Proc. 4th Int. Clean Air Congr., Tokyo, May, 1977.
21. **Meyer, H., Riemsloh, Z., and Krause, U.,** in *Aufbereitungstechnik,* 16, 245, 1975.
22. **Helfritch, D. J.,** Performance of an electrostatically aided fabric filter, *Chem. Eng. Prog.,* August, 54, 1977.

Chapter 4

ELECTROSTATIC PRECIPITATORS

I. INTRODUCTION

The electrostatic precipitator is used for the collection of fumes or fine solid particles (dust) suspended in the gas stream or in the exhaust pipe by the direct current-high voltage electrical field.[1-3] As shown in Figure 1, there is a collecting electrode formed as a cylindrical pipe to earth or of parallel plates to earth. At the center of a cylindrical pipe or at the middle of the plates, a fine wire, called a discharge electrode, is suspended by a weight. Then the negative voltage of 30 kV to 70 kV, obtained by a rectifier set from the alternating current, is added to the discharge electrode. Therefore, a corona discharge from the surface of the discharge electrode is produced in the electrical field.

At the same time, the dust-laden gas or fumes flow between both electrodes. Consequently, negative charged fine solid particles or negative charged fumes, are driven away by the Coulomb force to the surface of the collecting electrode and are collected on the surface. Increasing the thickness of the dust layer on this electrode with a lapse of time, this dust layer falls down into the dust bunker by a sharp impact, or rap in the dry-type precipitators. Clean air escapes from this electrostatic precipitator to the atmosphere. This is a fundamental physical process of the separation of the fine solid particles or fumes by electrostatic precipitators.

II. BRIEF HISTORICAL DEVELOPMENT

Hohlfeld (1824), a professor of mathematics in Leipzig, published the experimental results of the sedimentation of tobacco smoke by means of the discharge with a needle in a glass flask.[4] Lodge (1885) in England succeeded the sedimentation of magnesium oxide combustion smoke and of paper by a method similar to Professor Hohlfeld.[10] Lodge and industrialist A. O. Chester-Walker (1886) succeeded in separating mineral dust and smoke in an exhaust pipe by a small electrical chamber compared with a very large mechanical sedimentation chamber. Dr. K. Möller (1884) investigated the promotion of the sedimentation of fine solid particles by means of a high voltage discharge between a metal rod and a needle probe. Hempel (1886) had made a precipitation chamber, as shown in Figure 2, and separated the exhausted smoke. Cottrell (U.S.A.) and E. Möller (Germany) developed the modern electrostatic precipitator for industrial use. Cottrell (1911) applied the conception of Professor Lodge and solved the smoke pollution problem in the refinery equipment. Also E. Möller, the son of Dr. K. Möller, replaced the discharge electrode with a high tension fine wire and made the modern corona tube.

III. FUNDAMENTAL CONCEPTION OF ELECTROSTATIC PRECIPITATION

The fundamental construction of the electrostatic precipitators is composed of the discharge electrode generating a corona current and the collecting electrode building up the negative charged solid particles or negative fumes. Adding the high voltage to the discharge electrode, this electrode must be electrically insulated from the collecting electrode or casing by an insulator. This discharge electrode consists of a fine wire. The corona current produced by this electrode flows to the collecting electrode. Then the fine solid particles or the fumes flowing between both electrodes are charged by this corona current and move to the surface of the collecting electrode by Coulomb force and by the intensity of the electric field at the particle position.

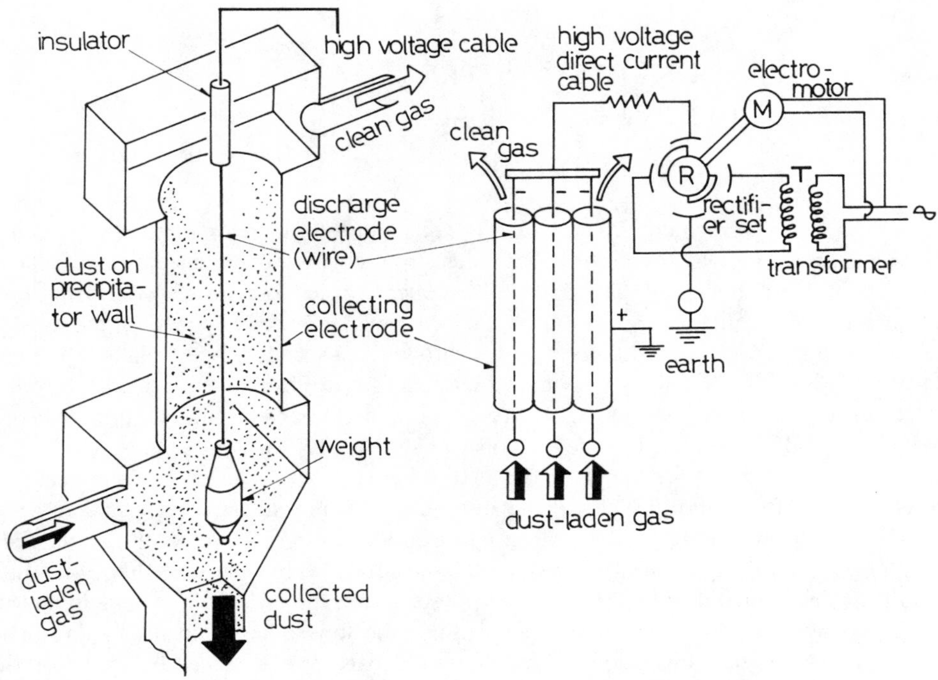

FIGURE 1. Illustrations of the electrostatic precipitator for general industry.

Consequently, the negative charged particles reaching the collecting electrode build some thickness of very tight dust layer. This dust layer separates and falls down from the collecting electrode to the dust bunker by the sharp impacts of the hammers. This physical mechanism is repeated and the dust collection takes place. Figure 3 represents the detailed mechanism of electrostatic precipitator collection of fine solid particles or fumes. Figure 3(a) represents the fine solid particle subjected to a charged state by the corona current from the discharge electrode. Figure 3(b) represents negative charged solid particles moving to the collecting electrode by the Coulomb force in the electric field. Figure 3(c) represents the dust layer building up on the surface of the collecting electrode and falling down with a hammer impact into the dust bunker by its own weight.

This is a fundamental concept of the electrostatic precipitation, but from the practical point of view, it is a very complicated collection process due to the combination of the gas properties and solid particle or fume properties.[5-7]

IV. FUNDAMENTAL TYPES OF ELECTROSTATIC PRECIPITATORS

The construction of electrostatic precipitators is divided into two systems. One system has a single-stage and the other is two stages, as shown in Figure 4. Figure 4(a) shows the single stage precipitator which realizes the charge and collection process for fine solid particles or fumes in the same section. This type of electrostatic precipitator is also applied in the collection of smoke or fumes in general industry. Figure 4(b) shows the two stage precipitator, which is divided into a particle charge section and a particle collection section. This type of the precipitator is applied in gas cleaning in ordinary building and also in the separation of aerosols in concentrations of several hundred milligrams per cubic meters of air. In the case of one-stage precipitator, dust concentration, dust property, and the flow rate of gas sometimes vary violently. Therefore, there are some problems of dust and gas

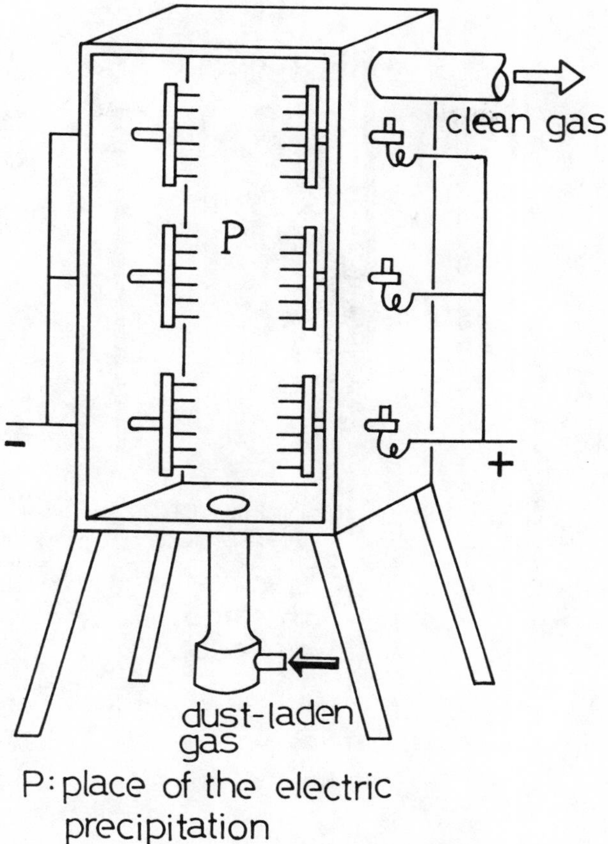

FIGURE 2. Electrostatic precipitator by Hempel (1886).

treatments in the one-stage precipitator. In order to make a high intensity electric field and constantly supply a stable corona current in the electrodes of negative characteristics in the one stage precipitator, a high voltage rectifier set is required. However, the space in this precipitator is small due to dealing with the charge and collection processes in the same section.

In order to describe the electrostatic precipitators for general industry, the characteristics of the one-stage precipitators are illustrated. The precipitators are divided into two types (cylinder type and plate type) of collecting electrodes, as shown in Figure 5. Figure 5(a) shows the cylinder-type precipitator which is composed of a cylindrical collecting electrode and a discharge electrode (wire) at the center of the cylindrical electrode. The dust-laden gas enters into this cylindrical collecting electrode and rises upward. This precipitator is sometimes applied as a wet precipitator of high collection efficiency $\eta_c = 99.9\%$. Figure 5(b) shows the plate-type precipitator which is composed of parallel plates of collecting electrodes and of discharge electrodes at the center between the plates. The dust-laden gas enters the collecting electrode plates. Most industrial precipitators are plate-type for the treatment of a huge flow rate of the dust-laden gas and reach the collection efficiency $\eta_c = 95\%$ for the dry type.

V. MOBILITY OF THE SOLID PARTICLE

Figure 6 shows charged solid particles moving in an electrical field. This problem is found

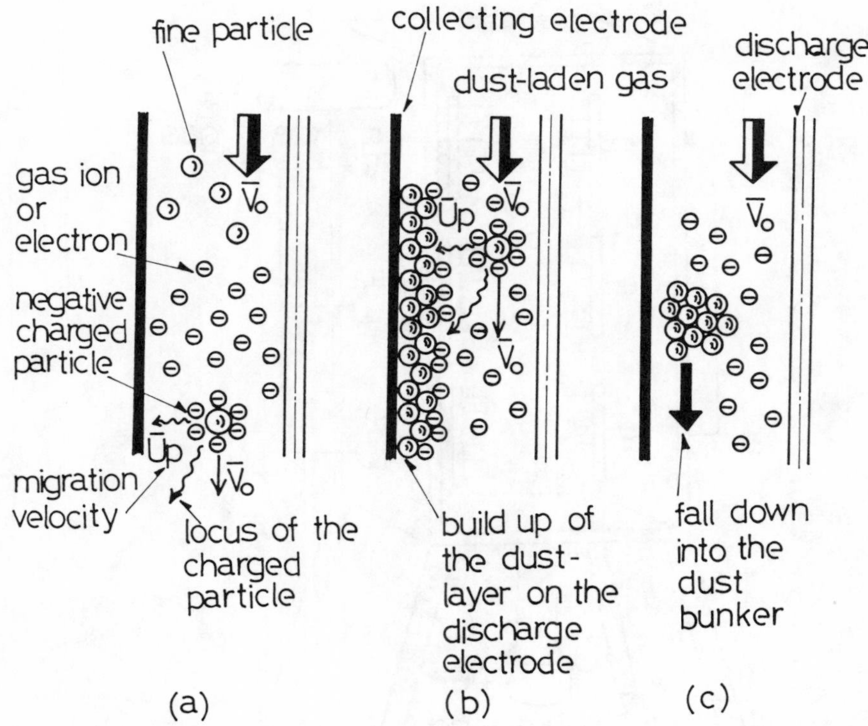

FIGURE 3. Fundamental physical mechanism of the collection of fine solid particles in the electrostatic precipitator.

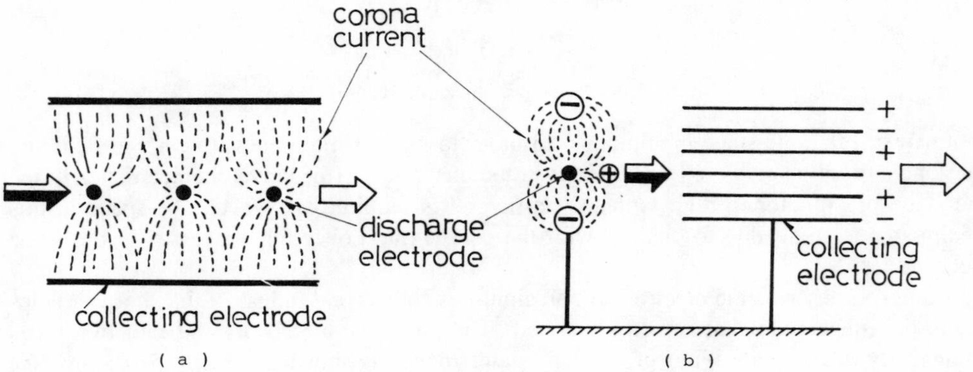

FIGURE 4. Illustration of the (a) one-stage precipitator of the parallel plate type and (b) two-stage precipitator.

increasingly in industrial applications of air conditioning, dust and fume removal, oil cleaning, and ore recovery by electrical processes.

Millikan[8] long ago showed that small charged particles (spherical solid particles) moving in a quiet gas under a field or under gravity force obey Stokes law. This law says that the force Fs(N) required to give a solid particle a velocity Up in the presence of the viscous drag can be written

$$Fs = 3 \cdot \pi \cdot \eta \cdot Xp \cdot Up \qquad (1)$$

where Xp is its diameter and η is the coefficient of gas viscosity. Now if a Coulomb force is $Fe = q \cdot E$ (N) due to the intensity of the electric field E(V/cm) acting on the electric

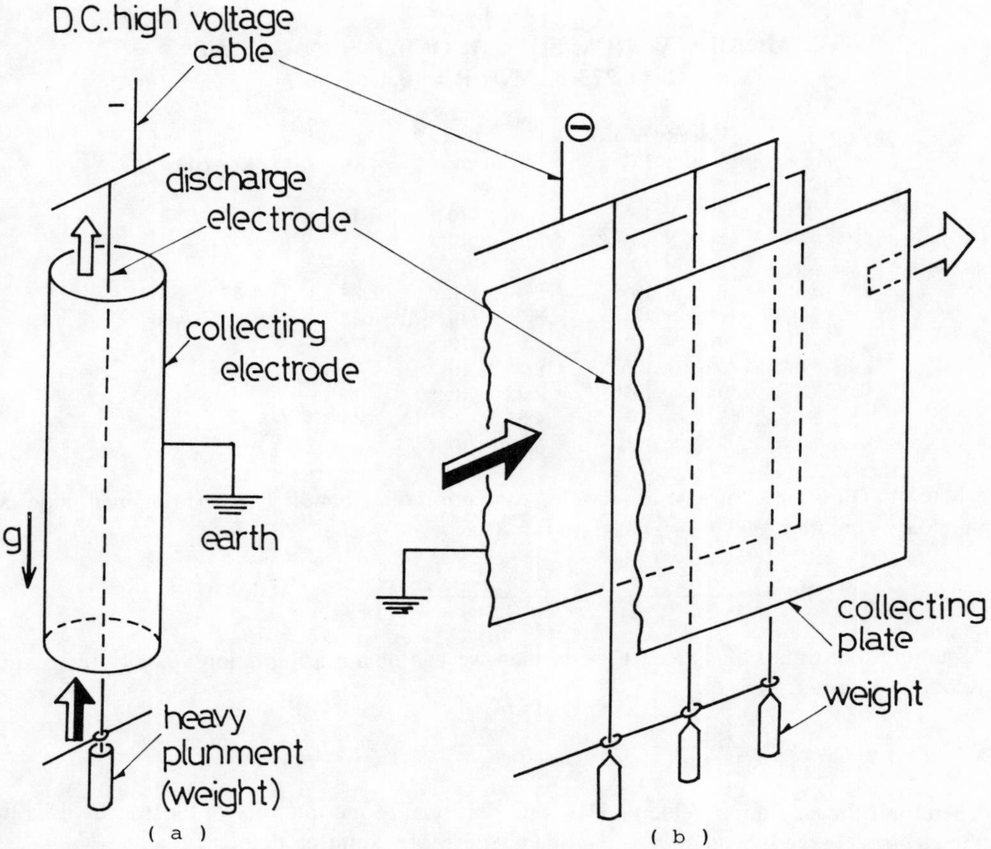

FIGURE 5. Fundamental types of the electrodes: (a) cylinder-type precipitator and (b) plate-type precipitator.

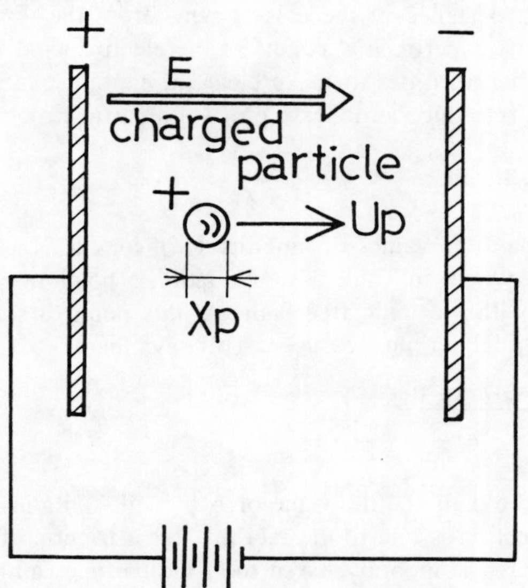

FIGURE 6. Electric field and charged solid particle.

charge q(C), then we can obtain a differential equation of solid particle motion in the quiet gas as follows:

$$mp\,\frac{dUp}{dt} = Fe - Fs = q \cdot E - 3 \cdot \pi \cdot \eta \cdot Xp \cdot Up \qquad (2)$$

Table 1
MOBILITY K(CM/S/V/CM) OF IONS IN GAS AT T = 273 K AND P = 0.1 MPA

Gas	Velocity (cm/s) at 273 K	λ(cm)	K(-ion)	K(+ion)
Air	4.83×10^4	9.9×10^{-6}	2.11	1.32
H_2	18.39×10^4	18.3×10^{-6}	8.15	5.92
O_2	4.61×10^4	9.95×10^{-6}	1.84	1.32
N_2	4.93×10^4	9.44×10^{-6}	1.84	1.28
He	13.11×10^4	28.5×10^{-6}	6.32	5.14
A	4.13×10^4	10.0×10^{-6}	1.71	1.32
Ne	5.61×10^4	19.3×10^{-6}	—	9.87
Cl_2	3.06×10^4	2.75×10^{-6}	0.74	0.74
SO_2	3.22×10^4	4.57×10^{-6}	0.407	0.407
k0NH_3	6.28×10^4	6.95×10^{-6}	0.658	0.565

where m*p* is the mass of a solid particle. A solution of Equation 2 can be obtained under a boundary condition of Up = 0 at time t = 0 as

$$Up = \frac{q \cdot E}{3 \cdot \pi \cdot \eta \cdot Xp}\left\{1 - \exp\left(-\frac{3 \cdot \pi \cdot \eta \cdot Xp}{mp} \cdot t\right)\right\} \tag{3}$$

Assuming that time t elapses as $t \rightarrow \infty$, then we can obtain a migration velocity (constant velocity) $\overline{U}p$(m/s) as

$$\overline{U}p = \frac{q \cdot E}{3 \cdot \pi \cdot \eta \cdot Xp} \tag{4}$$

Therefore, the migration velocity $\overline{U}p$ is proportional to the intensity of electric field E, to the electric charge q, and also to the diameter Xp of a solid particle.

From the practical point of view, the practical migration velocity is faster than $\overline{U}p$ of Equation 4 due to the existence of the corona wind from the discharge electrode to the collecting electrode. The experimental results of the electric wind will be described at the end of this chapter. Then in order to make clear the motion of fine solid particles in an electric field, we must introduce an index Kp of solid particle mobility as

$$Kp(\mathrm{cm/s/V/cm}) = \frac{\overline{U}p}{E} = \frac{q}{3 \cdot \pi \cdot \eta \cdot Xp} \tag{5}$$

Now in order to compare the values of mobility K of ions to Kp, some numerical values of the mobility K are written in Table 1.[9,10] Equation 5 holds until Xp becomes so small that it is comparable with the ionic free path. At this point, a semi-empirical correction factor Cm due to Cunningham must be used. This says that

$$Fs = \frac{3 \cdot \pi \cdot \eta \cdot Xp \cdot Up}{1 + A \cdot \dfrac{\lambda}{Xp}} \tag{6}$$

where λ is the mean free path and the value of A by Millikan's investigation is a constant value 1.748. The correction is due to the fact that for a friction of its path the particle is moving freely. Therefore the mobility Kp of the solid particle can be written

$$Kp = \frac{q}{3 \cdot \pi \cdot \eta \cdot Xp} \cdot \left(1 + 1.748 \cdot \frac{\lambda}{Xp}\right) \tag{7}$$

Now for the particles whose dimensions begin to be of the order of 10^{-5} cm at normal

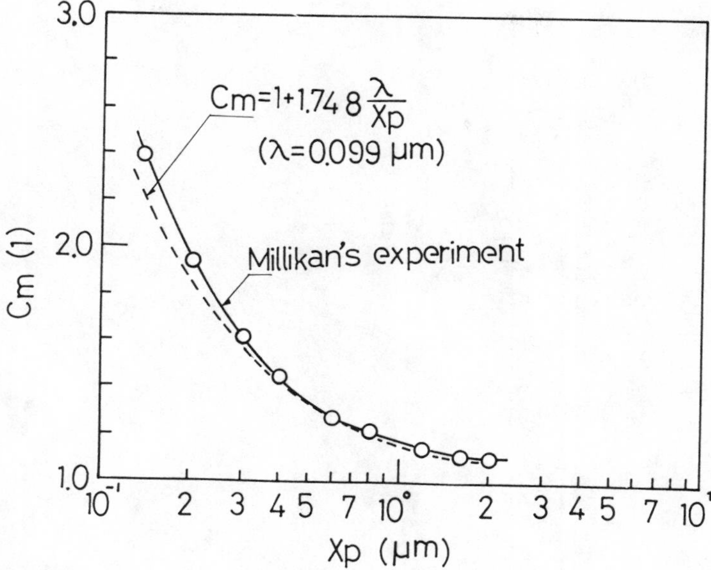

FIGURE 7. Cunningham's correction factor for a spherical particle in p = 0.1 MPa and T = 293 K.

condition, λ/Xp can no longer be neglected. As the particles become still smaller, 1.748 (λ/Xp) becomes much greater than 1 and the equation for Kp becomes

$$Kp = \frac{1.748\cdot\lambda\cdot q}{3\cdot\pi\cdot\eta\cdot Xp^2} \doteqdot 0.583\cdot\frac{\lambda\cdot q}{\pi\cdot\eta\cdot Xp^2} \tag{8}$$

Here, one numerical example is shown. Figure 7 shows the experimental results of Cunningham's correction factor[11] Cm by Millikan (1923) for a spherical particle in air (p = 0.1 MPa, T = 293 K) and the dotted line is an equation as

$$Cm = 1 + 1.748\cdot\frac{\lambda}{Xp}$$

where the value of λ is used as λ = 0.99 μm. The charge quantities q(C) on the fine solid particles[12] which are a function of the particle diameter Xp are shown in Figure 8. In order to evaluate the migration velocity $\overline{U}p$ (m/s) of fine solid particles in the electric field, the following numerical values are used: relative electric constant $\epsilon_S = 3$, intensity of electric field E = 3 kV/cm, electric current density of ion i_p = 0.2 mA/m², T = 300 K(air), p = 0.1 MPa, $\eta = 1.71 \times 10^{-5}$ (Pa·s), mobility(air) $K = 2.33 \times 10^{-4}$ m²/V·s, $\rho_p = 2 \times 10^3$ kg/m³, and terminal velocity in gravitational field $\underline{Wsg} = \rho_p\cdot g\cdot Xp^2/18\cdot\eta$.

From this figure we can see that the migration velocity $\overline{U}p$ of fine solid particles (Xp = 0.1 to 2 μm) in the electric field is a very high value in comparison with the terminal velocity Wsg in the gravitational field. This is a fundamental principle for an application of electrostatic precipitators separating fine solid particles and fumes from gas flow or air flow.

Figure 9 shows the experimental results of the apparent mobility Kp of fine solid particles (cement, Xp ≒ 2 μm; chloride, Xp ≒ 0.2 μm) of the diameter D1 on the cylindrical collecting electrodes; the diameter D2 of the discharge electrode; and the length L, where $\overline{E}p$ (kV/cm) means an average intensity of the electric field in a pipe. Those experimental values Kp were measured by Fukuda (1928)[13] and they show values comparable to Kp in Figure 8.

For more complicated electrode systems such as coplanar wires and parallel collecting electrode plates (plate type precipitators), even approximate calculations of the space-charge field become difficult and recourse must be made to experimental measurements. Measure-

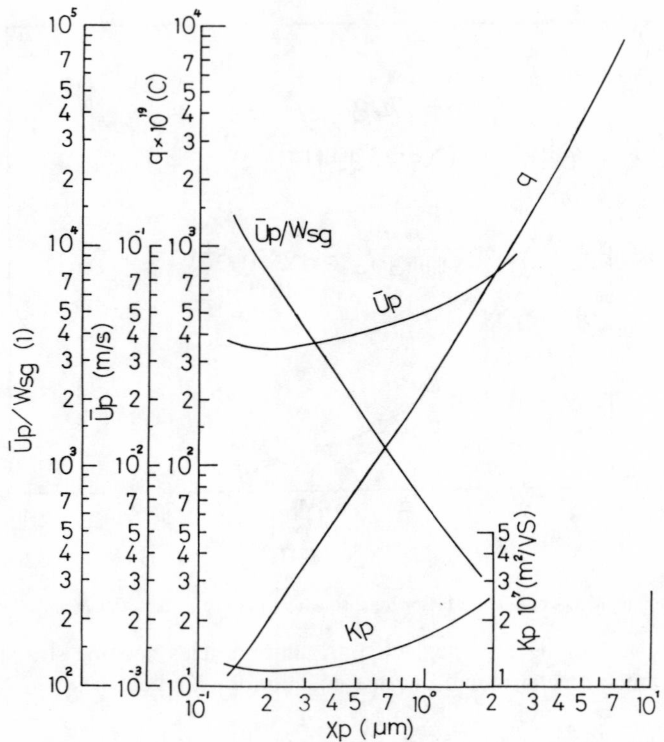

FIGURE 8. Charge quantity after one second under the following conditions; relative dielectric constant $\epsilon_s = 3$, $E = 3$ kV/cm, ic = 0.2 mA/m^2 (electric current density of ion), T = 300 K, p = 0.1 MPa (air), $K = 2.33 \times 10^{-4}$ m^2/V.s, $\rho_p = 2$/cm^3.

ments of this type were carried out in the laboratory of H. J. White[14] with a probe method as shown in graphical form in Figure 10. The electric field is seen to be much higher near the plate electrodes than would be the case if undistorted by the corona ions.

VI. PHYSICAL MECHANISM OF CORONA DISCHARGE

In general, in order to charge fine solid particles in the electrostatic precipitators, the characteristics of the corona discharge are applied.[15,16] There are positive and negative coronas on the corona discharge but negative coronas are more stable than are positive coronas. Therefore we can not only control the excess corona current but also get a more intense electric field. Consequently, characteristics of the negative corona can be applied in industrial electrostatic precipitators.

Now the author will describe the physical mechanism of the production of the negative corona. As shown in Figure 11, high voltage between the discharge electrode and the cylindrical collecting electrode is impressed, thus making the discharge electrode negative and the cylindrical collecting electrode positive. Then increasing the voltage, the intensity of the electric field E near the surface of the discharge electrode is high and E near the surface of the collecting electrode is low. When the voltage impress gives a low value, the state of gas or air in the electrodes is insulated. By increasing voltage impress between the electrodes, the intensity of the electric field also increases. Then electric current between the electrodes begins to flow. In this state, the electrons in gas or in air receive kinetic energy by the electric field and collide with the neutral molecules of gas. At this moment, as a result of the electrons giving its energy to the molecules, one neutral molecule becomes a positive ion by the separation of its election. This is a phenomenon of the ionization and

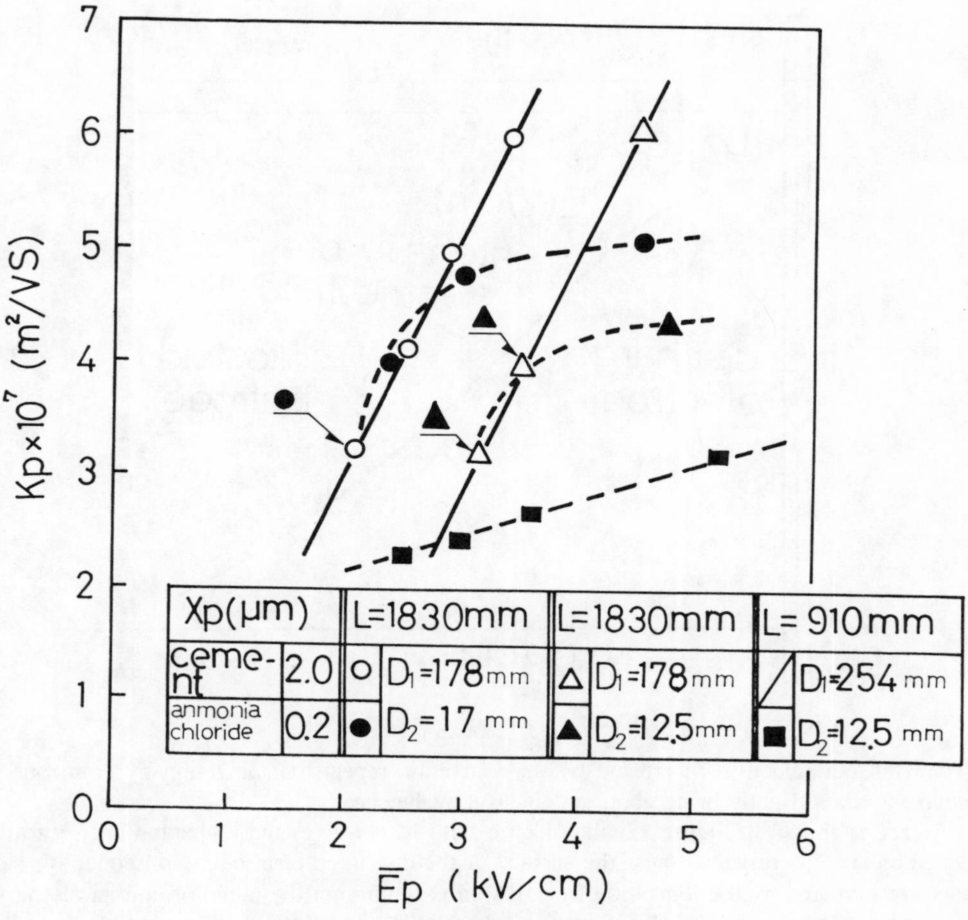

Kp×10⁷ (m²/VS)
7
6
5
4
3
2
1
0
0
1
2
3
4
5
6
Ēp (kV/cm)
Xp(μm)
L=1830mm
L=1830mm
L= 910mm
ceme-nt
2.0
○ D1=178mm
△ D1=178mm
D1=254 mm
anmonia chloride
0.2
● D2=17 mm
▲ D2=12.5mm
■ D2=12.5 mm

FIGURE 9. Average intensity of electric field in a pipe.

collecting electrode
V=0kV
8
A
B
discharge electrode
V=40kV
16
24
32
40kV
V=40 kV
149
200 mm
599 mm
A
B
40 30 20 10 0
V (kV)

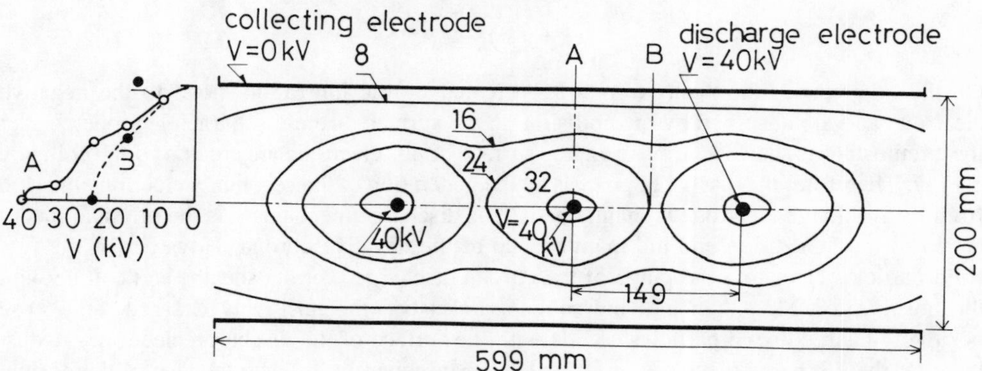

FIGURE 10. Equi-potential lines between discharge and collecting electrodes in duct-type electrostatic precipitator.

the minimum energy required to excite the ionization of the neutral molecules is called the ionization voltage. The voltage impress between the electrodes increases more than the ionization voltage, so the neutral molecules M ionize one after another and become the positive ion (M^+) and electron e,

$$e \rightsquigarrow \underset{\text{neutral molecule}}{\textcircled{M}} \longrightarrow \underset{\text{positive ion}}{M^+} + \underset{\text{electron}}{e}$$

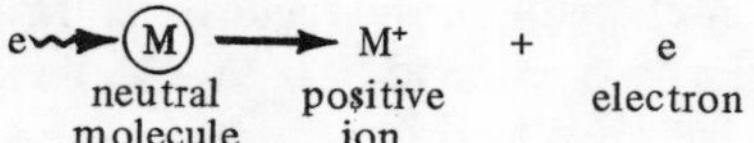

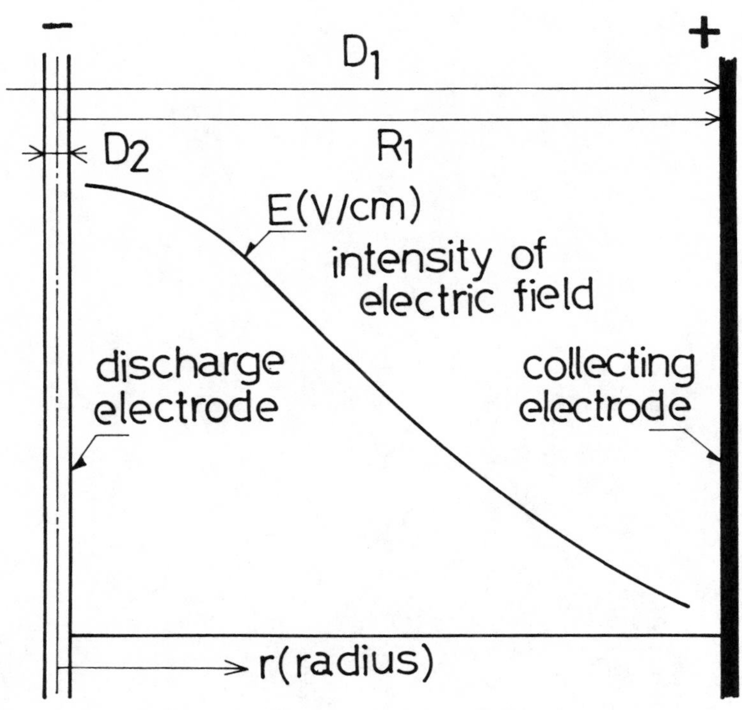

FIGURE 11. Distribution of electric field.

Then the new electrons e created by the ionization repeat the ionization by collisions in succession and finally bring about an electron avalanche.

Here, as shown in Figure 12, the electric field which gives the ionization energy to the gas molecules is restricted near the surface of the discharge electrode. Consequently, new electrons created by the ionization by collision pass through the range of corona discharge and move to the positive collecting electrode, and then collide with the neutral molecules M. At this moment, the neutral molecule becomes a negative ion −M as

$$e + (M) \longrightarrow M^{-}$$

On the other hand, the positive ions M^{+} created by the ionization move to the negative electrode and are absorbed by the collision on the surface of the discharge electrode. Under the circumstances, new electrons from the surface of this electrode are created by the collision energy. Therefore, the discharge process by the generation of the secondary electron emission from the discharge electrode is maintained. This discharge mechanism is the physical mechanism of the generation and the maintenance of the negative corona. However, if the fine solid particles enter into the range of the corona discharge (corona sheath) and collide with the positive ion M^{+}, then the fine solid particles become positively charged, and those positively charged solid particles collide with the surface of the discharge electrode. A dust layer on the discharge electrode is formed and consequently the maintenance of the normal corona discharge becomes impossible.

The radius Rc of the corona sheath can be estimated by the equation of Fazel and Parsons (1924)[17] as

$$Rc = R2 + \left(1.5 - 0.35\right) \times 10^{-5} \times (E - Ec)$$

where R2 is a radius of the discharge electrode and Ec is the corona starting voltage. For example, if R2 = 0.05 cm, Ec = 5 kV, and E = 90 kV, then the radius Rc of the corona

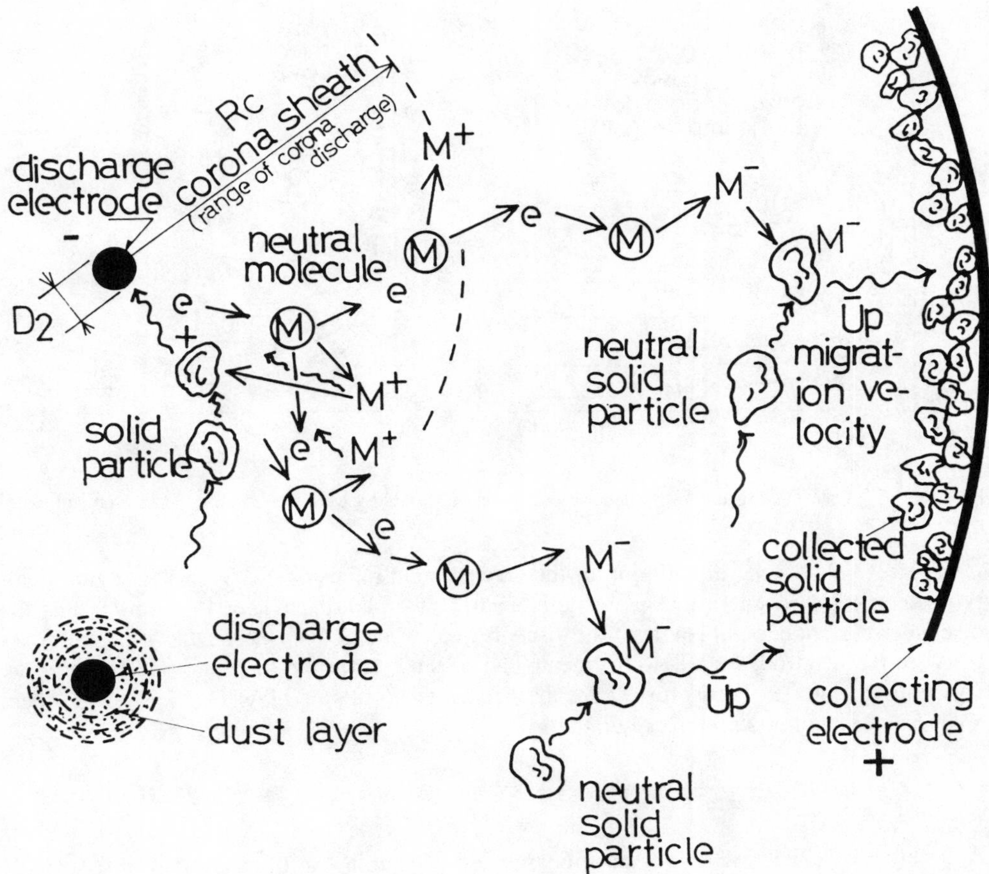

FIGURE 12. Illustration of the charge mechanism by the negative corona.

Table 2
THE INTENSITY OF THE CORONA STARTING ELECTRIC FIELD (STANDARD CONDITION)

Gas	Intensity of the corona starting electric field (kV/cm)
Air	35.5
H_2	15.5
O_2	29.1
N_2	38.0
CO	45.5
CO_2	26.2
NH_3	22.3

sheath is Rc = 1.32 cm, and also R2 = 0.6 cm, Ec = 25 kV, then R2 = 1.58 cm. Here, the intensity of the corona starting electric field is written for the standard condition in Table 2.

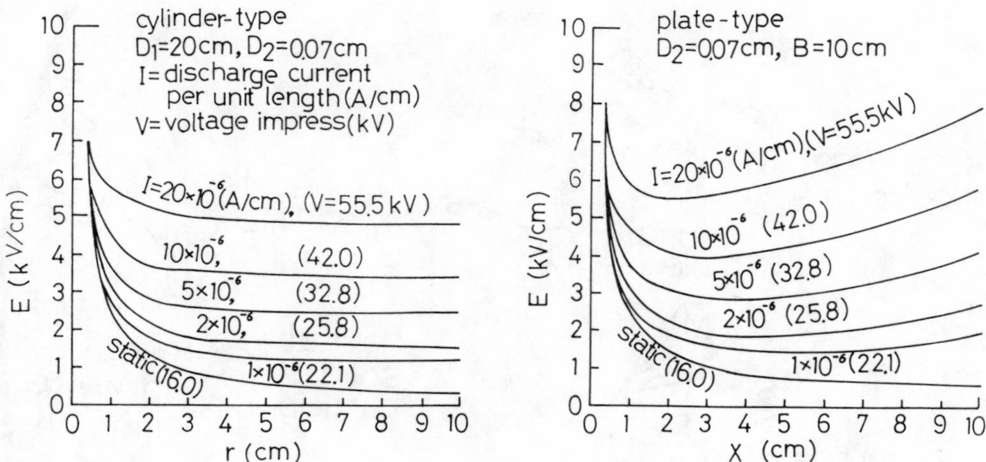

FIGURE 13. Calculated results of the intensity of the electric field by Mayr (1927) in cylinder-type and plate-type electrostatic precipitators.

Figure 13 shows the distribution of the electric field intensity E (kV/cm) for cylindrical-type electric precipitators and plate type electric precipitators, where I (A/cm) means the discharge-current per unit length of the discharge electrode, V(kV) means the voltage impress between the electrodes, and X(cm) means the distance from the discharge electrode to the collecting electrode. Those curves are the calculated results of Mayr (1927)[18] and in them a curve of a symbol static is calculated by

$$E(r) = \frac{V}{r \ln (R1/R2)} = \frac{2.83}{r} \text{(kV/cm)} \tag{9}$$

As is known in the case of no-space charge, this equation can be derived from Laplace's differential equation

$$\nabla^2 V = \frac{1}{r}\cdot\frac{\partial}{\partial r}\left(r\cdot\frac{\partial V}{\partial r}\right) + \frac{1}{r^2}\cdot\frac{\partial^2 V}{\partial \theta^2} + \frac{\partial^2 V}{\partial z^2} = 0\left(\frac{V}{m^2}\right) \tag{10}$$

However, in the case of the existence of the space charge between the electrodes by the generation of the corona discharge, the intensity of the electric field can be determined by solving Poisson's equation

$$\nabla^2 V = -\frac{\rho}{\epsilon_0}\left(\frac{V}{m^2}\right) \tag{11}$$

where $\epsilon_o = 8.854 \times 10^{-12}$ (F/m) means a dielectric constant and ρ(C/m^3) means the volumetric density of charge. Then the solution of this equation can be written

$$E(r) = \sqrt{\frac{2I}{K} + \left(\frac{Ec\ R2}{r}\right)^2} \tag{12}$$

where I(A/m) means the corona current per unit length of the discharge electrode, K(m^2/V·s) means the mobility of ion, Ec (kV/m) means corona starting strength of electric field, and R2(m) means the radius of the discharge electrode.

Figure 14 shows the experimental results of the intensity of the electric field with parallel plate electrodes with a point electrode inserted at the center of one of the plates.[19] Similar results occur with wire-cylinder systems.

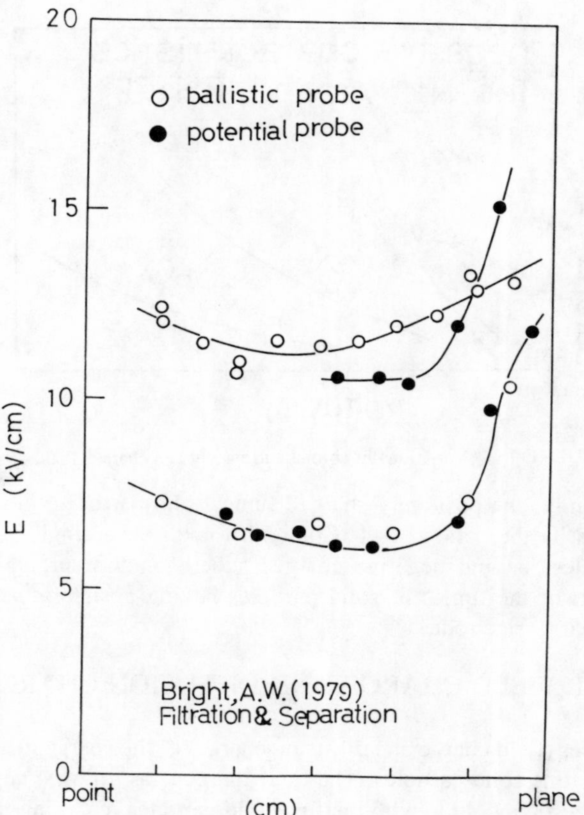

FIGURE 14. Distribution of the electric field in point-plane geometry.

VII. MECHANISM OF CHARGE ON SOLID PARTICLES

In normal air, unipolar ion concentrations under electrostatic precipitation commonly range from 10^8 to 10^9 ions per cm^3, corresponding to the strength of the electric field from 1 to 10 kV/cm. Corona current-voltage characteristics are a function of gas composition and in practice vary widely depending on the nature of the gas.[20]

The electrons passing through the corona sheath do not have enough energy for ionization of gas molecules due to the weak electric field intensity. In the weak intensity space of an electric field out of the corona sheath, negative ions which are generated by the adhesion of electrons on the neutral gas molecules are governed, but the probability of electron adhesion depends on the kinds of gas molecules. Figure 15 shows the corona-current voltage characteristics for the typical gases N_2, O_2, and $CH_3{\cdot}Cl$.[14]

On the electronegative gas (for example, gas containing molecule construction with O_2 or containing halogen), the electrons generated by the electron avalanche collide with the neutral molecules of gas, then they adhere with high probability. The electron adhesive neutral molecule becomes a negative ion. The mobility Kp of the negative ion produced by the above stated process is too slow in comparison with that K of the ions. Therefore the fume, or fine solid particles, entering the space charge domain of the negative ions between the electrodes become the negatively charged particles by the adhesion of negative ions.

Then the negatively charged particles move to the positive collecting electrode by the action of Coulomb force in the electric field between the electrodes. On the other hand, a part of the positive ions generated by the ionization collides with the fume or fine solid

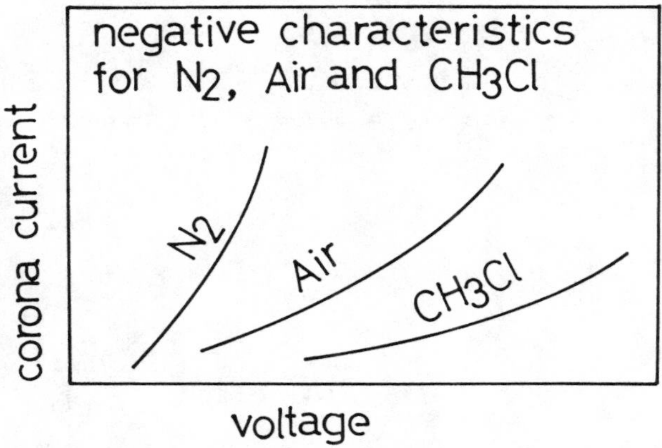

FIGURE 15. Variations in corona-current voltage characteristics.

particles. Consequently, the positively charged fume, solid particles, move to the negative discharge electrodes by the action of the Coulomb force. Accordingly most of the fumes, or fine solid particles, having negative charges adhere on the surface of the collecting electrode, but a part of the fumes or solid particles having positive charges adhere to the surface of the discharge electrode.

VIII. FIELD CHARGE AND DIFFUSION CHARGE

There are two types (field charge and diffusion charge) of the charge mechanisms adhering negative ions on the fine solid particles. The field charge is as follows: negative ions having energy by the electric field collide with the fine solid particles, charging them. On the other hand, diffusion charge is as follows: due to the irregular thermal ion motion by the law of the kinetic theory of gases, ions adhere on the solid particles by the diffusion effect and the solid particles are charged.

In electrostatic precipitators, the charge process of fine solid particles is governed by both charge mechanisms due to the existence of the electric field. Figure 16 shows the relationship between the number n of the electron charge and the particle diameter Xp.[21] Then if the particle diameter Xp is less than Xp = 0.5 μm, the diffusion charge is governed. However, if the particle diameter Xp is larger than Xp = 0.5 μm, then the field charge is governed. Figures 17 and 18 show the experimental results of the charging process related to the particle diameter Xp by H. J. White (1951), where n is the number of charges.

IX. DISCHARGE CHARACTERISTICS

A. Corona Onset Field Strength

Increasing voltage V between the discharge electrode and collecting electrode, corona discharge on the discharge electrode occurs. The voltage and strength of the electric field are called the corona starting voltage Vc and corona onset field strength Ec. Those values are estimated as follows.

For cylinder type of the electrodes, denoting R2 as the radius of the discharge electrode and R1 as the radius of the collecting electrode, the intensity of the electric field E can be written from the Gauss theorem as

$$E(r) = \frac{V}{r \cdot \ln(R1/R2)} \tag{13}$$

where V is the voltage between the electrodes, r is the radius distance from the discharge electrode, and E(r) is the intensity of electric field at radius r. This equation is applied for

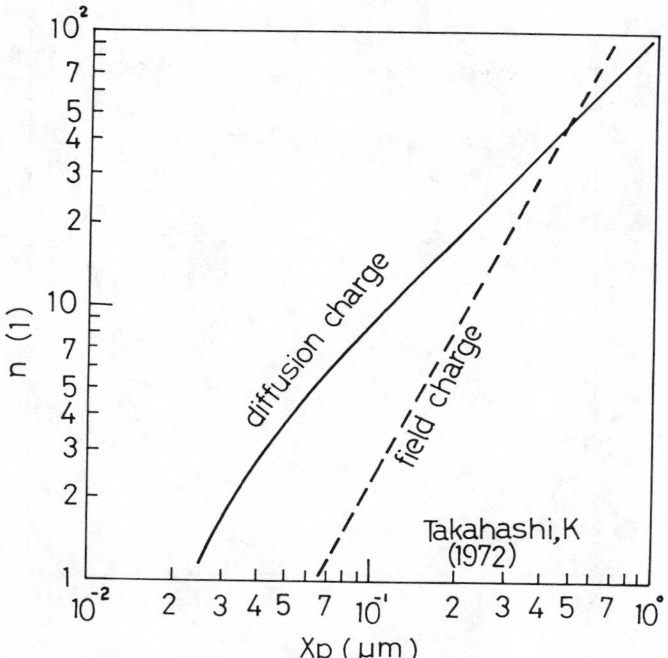

FIGURE 16. Relationship between number of electron charge and particle diameter.

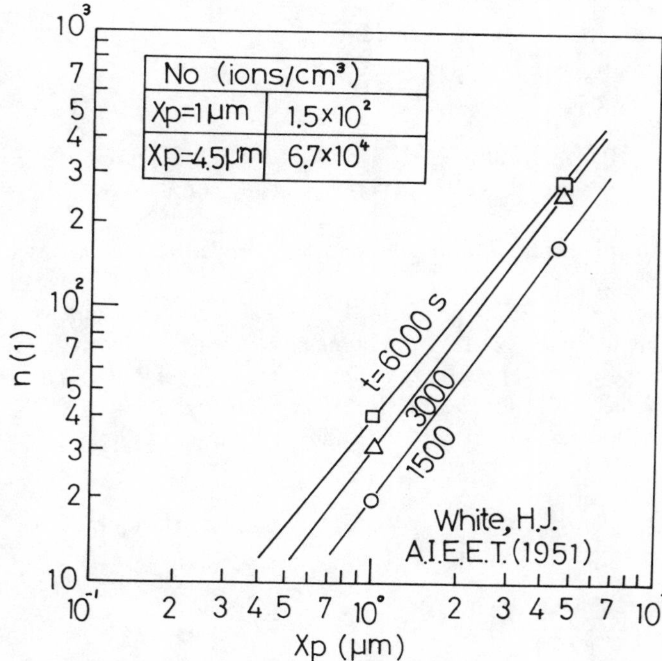

FIGURE 17. Experimental results of diffusion charging process.

the state before the corona starting voltage Vc. Increasing voltage V, the corona discharge occurs.

The corona onset field strength Ec for the negative corona can be written

$$Ec = \left(31.03 \cdot \delta + 9.54 \cdot \sqrt{\frac{\delta}{R2}}\right) \left(\text{kV/cm}\right) \tag{14}$$

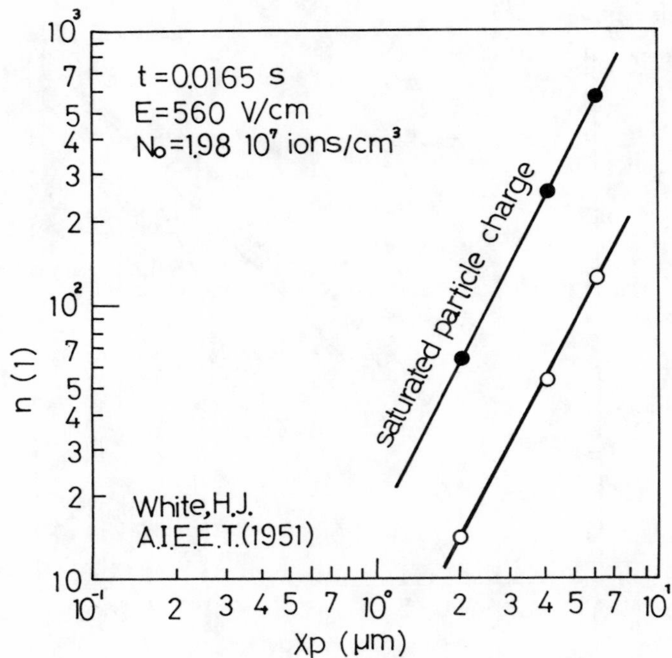

FIGURE 18. Experimental results of field charge process.

where δ is the relative density of air defined as

$$\delta = \frac{p}{10^5\,(\mathrm{Pa})} \times \frac{298}{273 + t} \tag{15}$$

where p is pressure (Pa) of air at temperature (273 + t)K and an unit of radius R2 is cm. From Equation 14, we can find the characteristics of the corona onset field as follows:

1. Decreasing the diameter D2 of the discharge electrode, the corona onset field strength Ec becomes higher.
2. Going away from the discharge electrode, the value of Ec becomes rapidly lower.
3. Increasing the diameter D2 of the discharge electrode, the value of Ec near the surface of the discharge electrode becomes lower. However, going away from the discharge electrode, the degree of the descendent of Ec becomes lower.
4. Increasing the temperature of gas, the value of Ec becomes lower.[22]

B. Corona Starting Voltage

The corona starting voltage Vc can be obtained by integrating Ec from the surface of the discharge electrode to the wall of the collecting electrode as

$$\mathrm{Vc} = \int_{R2}^{R1} E(r)\cdot dr = \int \frac{R2}{r}\cdot \mathrm{Ec}\cdot dr = R2\cdot \mathrm{Ec}\cdot \ln\frac{R1}{R2} =$$

$$= \left(31.03\cdot\delta + 9.54\cdot\sqrt{\frac{\delta}{R2}}\right)\cdot R2\cdot \ln\frac{R1}{R2}\cdot \tag{16}$$

where the units R1 and R2 are centimeters. Consequently, if the voltage increases up to Vc, the electrons near the discharge electrode can be accelerated until they have enough ionization energy. At that moment the electron avalanche is brought about. Then the corona

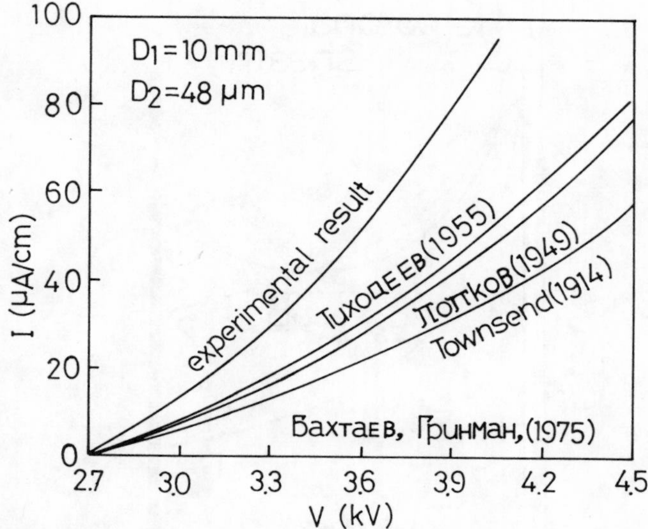

FIGURE 19. Relationship between voltage and corona current in the electrodes.

near the discharge electrode begins to appear and this voltage Vc is called the corona starting voltage. Therefore, the space charge by corona and the space charge by the charged solid particles have an influence on the strength of the electric field.

C. Characteristics of Corona Discharge

As shown in Figure 19, a relationship between V(V) and the corona current (A/m) in the electrodes is generally called a characteristic of corona discharge and is one of the most important characteristics for the electrostatic precipitators.[16]

There are many papers concerning the corona current in electrodes. But for low values of the corona current, Townsend (1914)[8,23] derived an equation of the corona current I as

$$I = \frac{V \cdot (V - Vc)}{9 \times 10^5} \cdot \frac{2K}{R1^2 \cdot \ln(R1/R2)}, \ (A/cm) \tag{17}$$

where K(cm^2/V·s) is ion-mobility and the units R1, R2 and V1, Vc are cm and kV, respectively. For high values of the corona current, the following approximate equation of the corona current I is written

$$I = \frac{2K}{R1^2} \cdot \left\{ (V - Vc)^2 + \frac{Vc \cdot (V - Vc)}{\ln(R1/R2)} \right\} \cdot \frac{1}{9 \times 10^5} \tag{18}$$

And Popkov (1949) gave the following equation

$$I = B \cdot V \cdot (V - Vc) \cdot \frac{2K}{R1^2 \cdot \ln(R2/R1)} \cdot \frac{1}{9 \times 10^5} \tag{19}$$

where a unit of I is A/cm and B is the constant value of the correction factor. On the other hand Tihodeev (1955)[24] gave the following equation

$$I = \frac{2KV(V - Vc)}{R1^2 \cdot \ln\frac{R1}{R2}} \cdot \left(\frac{(V - Vc)\ln(R1/R2)}{4V} + \frac{Vc}{2V}\left[1 + \left\{ 1 + \frac{2}{\left(\frac{V}{Vc} - 1\right) \cdot \ln\frac{R1}{R2}} \right\} \times \right.\right.$$

$$\left.\left. \times \ln\left\{ \frac{1}{2} \cdot \left(\frac{V}{Vc} - 1\right) \cdot \ln\frac{R1}{R2} + 1 \right\} \right] \right) \cdot \frac{1}{9 \times 10^3} \ (A/cm) \tag{20}$$

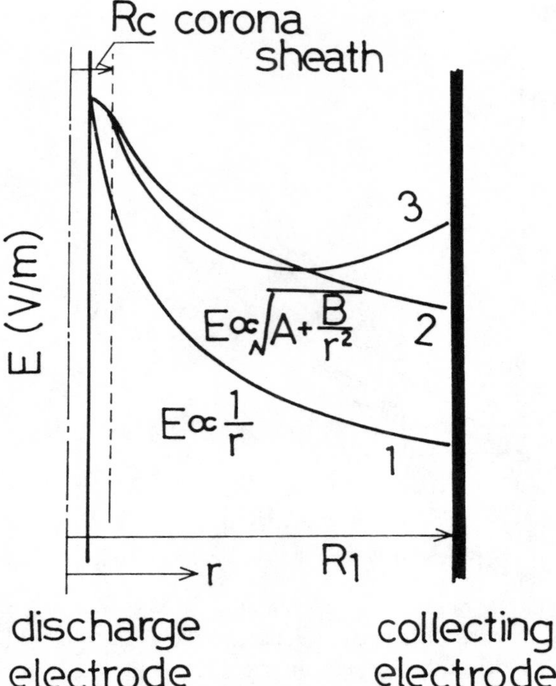

FIGURE 20. Typical distribution of the intensity of the electric field for the cylinder-type electrostatic precipitator.

where the units K, R1, and R2 and V and Vc are $cm^2/V{\cdot}s$, cm, and kV, respectively. From those equations, the corona current mainly depends on the ion mobility K.

X. DISTRIBUTION OF THE INTENSITY OF THE ELECTRIC FIELD

The characteristics of the intensity of the electric field strongly depend on the distribution of the space charge by the fine solid particles or fumes in electrostatic precipitators. Then the intensity of the electric field which is distributed in the electrodes is composed of the two factors as follows:

1. An electrostatic field which is controlled by the geometrical forms of the electrodes and by the voltage between the electrodes
2. The intensity of the electric field which is related to the space charge produced by the existence of the ions and the charged fine solid particles

An example of the cylinder type electrostatic precipitator, the typical distributions of the intensity of the electric field are shown in Figure 20. Curve 1 is a state of the no charges in the electrodes, which means no corona discharge exists. As stated above, the intensity of the electric field E(r) at radius r can be written

$$E(r) = \frac{V}{r \cdot \ln(R1/R2)} \tag{21}$$

Curve 2 is the state of existence of the ion space charge ions in the electrodes. The intensity of the electric field E(r) can be obtained by solving Poisson's equation

$$E(r) = \sqrt{\frac{2I}{K} + \left(\frac{Ec \cdot R2}{r}\right)^2} \tag{22}$$

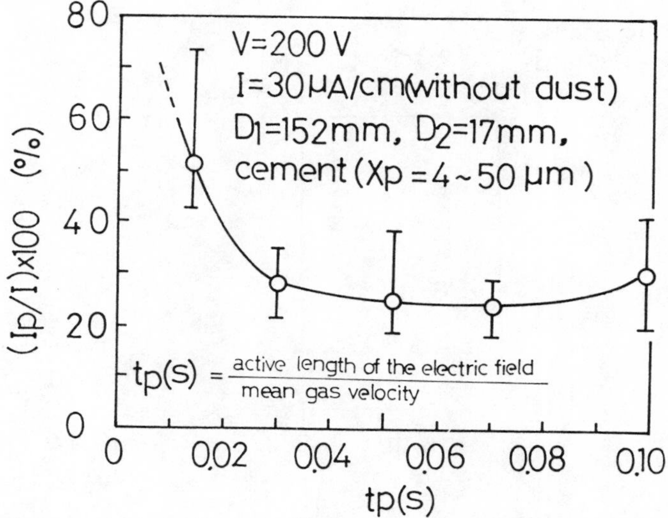

FIGURE 21. Experimental results of the charging-up process for cement dust particles.

Curve 3 is the state of the existence of the charged solid particles or of charged fumes which are distorted by the electric field in the electrodes, where K is the ion mobility. Near the surface of the collecting electrode, the value of 2I/K is very large in comparison with that of $(Ec \cdot R2/r)^2$, therefore Equation 22 which is independent of the radius r can be simplified to

$$E \fallingdotseq \sqrt{\frac{2I}{K}} \tag{23}$$

In addition to the geometrical conditions of the electrodes and gas characteristics, there are two important effects: influence of the space charge by the charged solid particles having very slow mobility Kp, and the influence of the discharge characteristics resulting from the adhesion of dust on the surface of the discharge electrode.

An experimental result by Fukuda (1928)[13] for the corona-current Ip on the dust-laden gas, which is related to the remained time t (s) of the solid particles moving in the electric field to the corona current I without dust, is shown in Figure 21. The experimental conditions were as follows: the voltage between the electrodes is 200 V, the corona current without the solid particles is I = 30 μA/cm and the mean dust (cement dust, Xp = 4 to 50 μm) concentration is Co ≒ 170 g/m^3, diameter of the collecting electrode is D1 = 152 mm, and of the discharge electrode is D2 = 17 mm, its active length is 305 mm, and the mean gas velocity is between 0.91 m/s and 21.3 m/s. From the experimental results of Figure 21, we find that the required time from when the fine solid particles enter the electric field to arrival at the sufficient saturation charge on the solid particles is very short.

XI. APPARENT RESISTIVITY

One of the most important factors for the fine solid particles and for the fumes is the apparent resistivity which controls the collection efficiency η_c and also affects the discharge characteristics for electrostatic precipitators. In general, the apparent resistivity is defined as follows: resistance is a proportional factor of Ohm's law having a value corresponding to the kinds of materials, forms, and temperature. Then, if the form of the materials and the temperature are assumed to be a constant value, the value of the resistance can be determined by the kinds of the materials. Therefore, forming a cube of the unit length, as shown in Figure 22, the resistance of this material is called a resistivity.

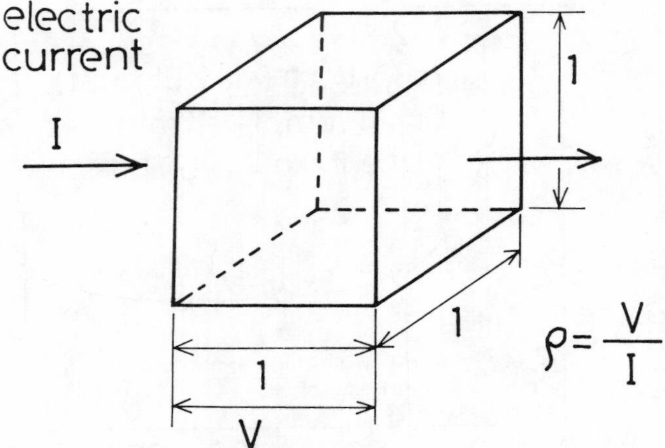

FIGURE 22. A cube for resistivity.

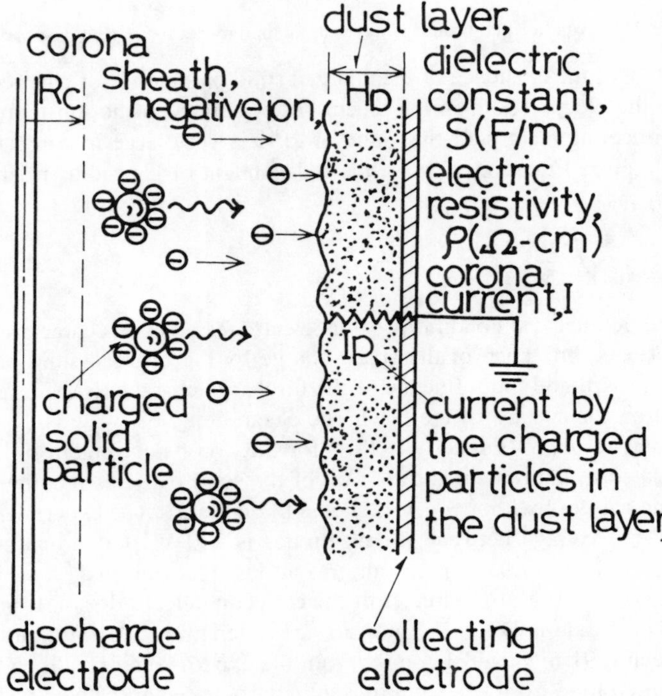

FIGURE 23. Illustration of charged particles and dust layer.

Consequently, in the case of fine solid particles in the electrostatic precipitator, charged solid particles are collected on the surface of the collecting electrode to form the dust layer. Then under the conditions of the given gas temperature and relative humidity, the resistivity which is converted to the resistivity of the above defined cube is called an apparent resistivity, which is different from the ordinary resistivity.

As shown in Figure 23, charged solid particles having a mobility Kp collide with the collecting electrode and form the dust layer. The thickness H_D of this dust layer increases with time by colliding the charged particles one after another with the dust layer. Many of the electrons which do not collide with solid particles directly give their charge to the dust layer. Therefore, the corona current I is composed of the current Ip by charged solid particles and of the current Ii by the ions.

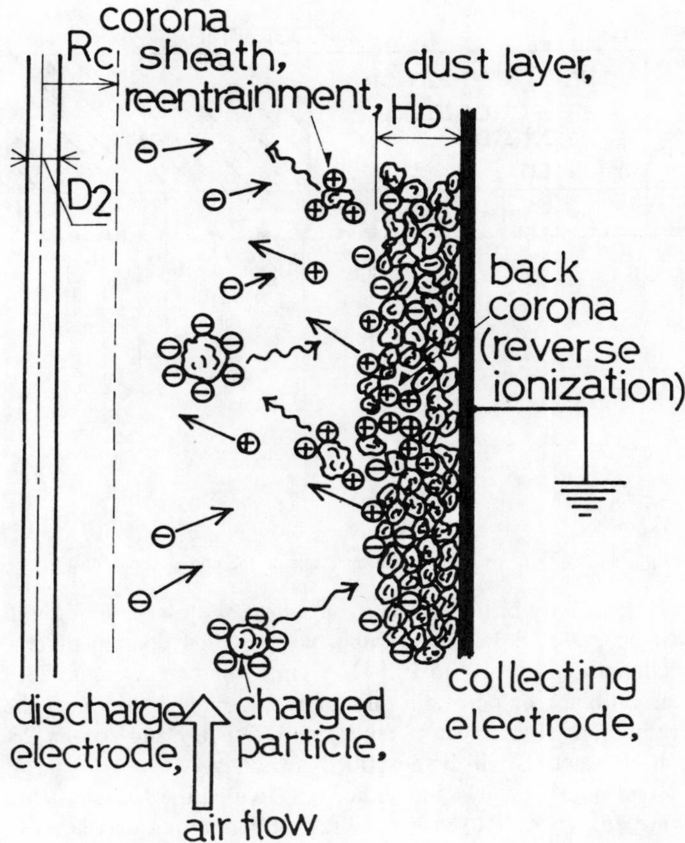

FIGURE 24. Physical explanation of back corona.

Now if charged solid particles collide with the surface of the collecting electrode, then those solid particles are supported on the collecting electrode by the Coulomb force. On the other hand, the charged quantity on the solid particles is gradually released depending on the resistivity of the solid particles or fumes. Therefore, the adhesion force on the collecting electrode is decreased with time. Accordingly, low resistivity fumes or particles ($\rho \leqq 10^3$ $\Omega - \text{cm}$) rapidly release the charges, and then the adhesive force by Coulomb force decreases very rapidly. As a result of this phenomenon, the solid particles on the collecting electrode are detached and again are returned to the gas stream. This phenomenon is called dust re-entrainment which sometimes occurs by treating carbon-black as a low resistivity particle in the electrostatic precipitator.

In the case of the high resistivity solid particles ($\rho \geqq 10^{12}$ $\Omega - \text{cm}$), many ions collide with the collected solid particles in the dust layer, thereupon the accepted charge quantity increases with time and the surface potential of the dust layer also increases. Increasing the surface potential in the free space of this dust layer, the dielectric breakdown occurs and, at the same time, new ions are created. In those new ions, positive ions obey the Coulomb law and move to the discharge electrode from the surface of the dust layer. Then moving to the discharge electrode, positive ions lose their charges by colliding with the negative gas ions, or with charged solid particles.

In a case of high resistivity materials, the effect of collection efficiency is decreased. This phenomenon is called reverse ionization, or back corona. This kind of a phenomenon sometimes occurs in the electrostatic precipitator collecting the exhausted fly-ash from a boiler. Figure 24 shows the physical mechanism of the back corona.[25]

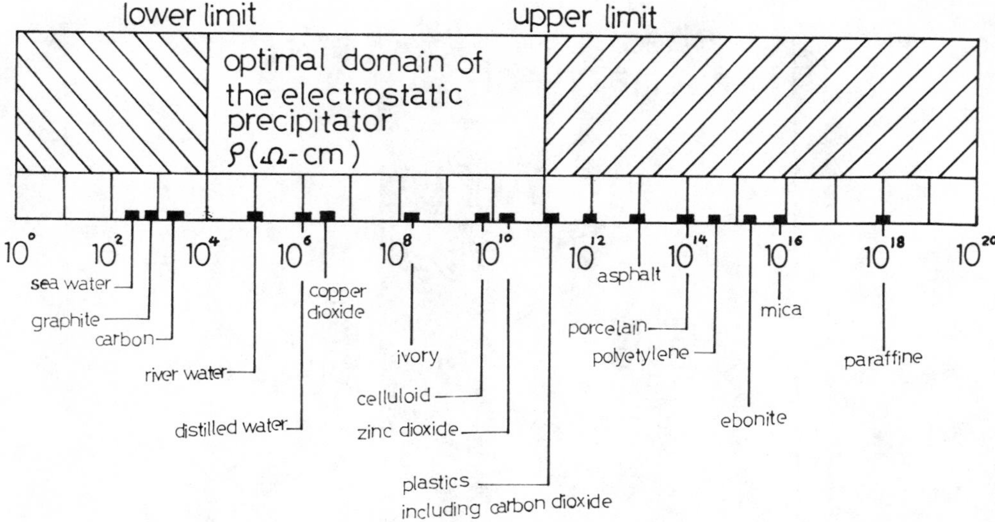

FIGURE 25. Resistivities of the materials on the normal air condition.

The property of resistivity of the material for collecting dust or fume by the electrostatic precipitator must be grasped. In general, the most optimal domain of resistivity for the electrostatic precipitator is $\rho = 10^4$ to 10^{11} $\Omega - $cm, as shown in Figure 25. Consequently, for collecting carbon-black or porcelain dust, it is not enough to collect the dust only by electrostatic precipitators. The effect of collection efficiency and corona discharge for resistivity variation of materials will be described.

From the physical point of view, the author has divided the four domains of resistivity, as shown in Figure 26, where I(%) means the rate of increase or decrease of corona current (see Hashimoto, K. and Taniguchi, A., *Theory and Practice of the Electrostatic Precipitators,* (in Japanese), Denki-Shoin, Japan, 1965).

Figure 27 shows the resistivity of fly-ash for humidity which was measured by Penney (1951).[26] From this figure, we find that the resistivity ρ is varied depending on the relative humidity.

On the other hand, the simplified result (Sproull and Nakada, 1951) represents the voltage V(V) across the thickness H_D of the dust layer when the electrical equilibrium is attained as $V = \rho I H$. Under this equilibrium condition, the current leaking through the dust layer H_D to the collecting electrode exactly balances the current $I = Ip + Ic$ arriving at the exposed surface of the dust layer due to the arrival of additional charged dust combined with the electronic current from the discharge electrode. Thus the expression for the potential drop essentially reduces to Ohm's law.

XII. REVERSE IONIZATION PHENOMENA

As described above, when the resistivity of dust or fumes is very high, the partial discharge inside the dust layer on the collecting electrode occurs. The positive ions within the gas, which are generated by the polarization, move to the discharge electrode from the surface of the dust layer, as shown in Figure 24.

For example, this partial discharge inside the fly-ash dust layer in the intensity of the dust layer electric field will happen to be $E_D = 10$ to 15 kV/cm. Then Figure 28 shows the intensity of the electric field across the dust layer H_D, which is related to the resistivity ρ.

Further, even if the fine solid particles or fumes are composed of the same form and chemical component, decreasing the particle diameters Xp, the resistivity is decreased due

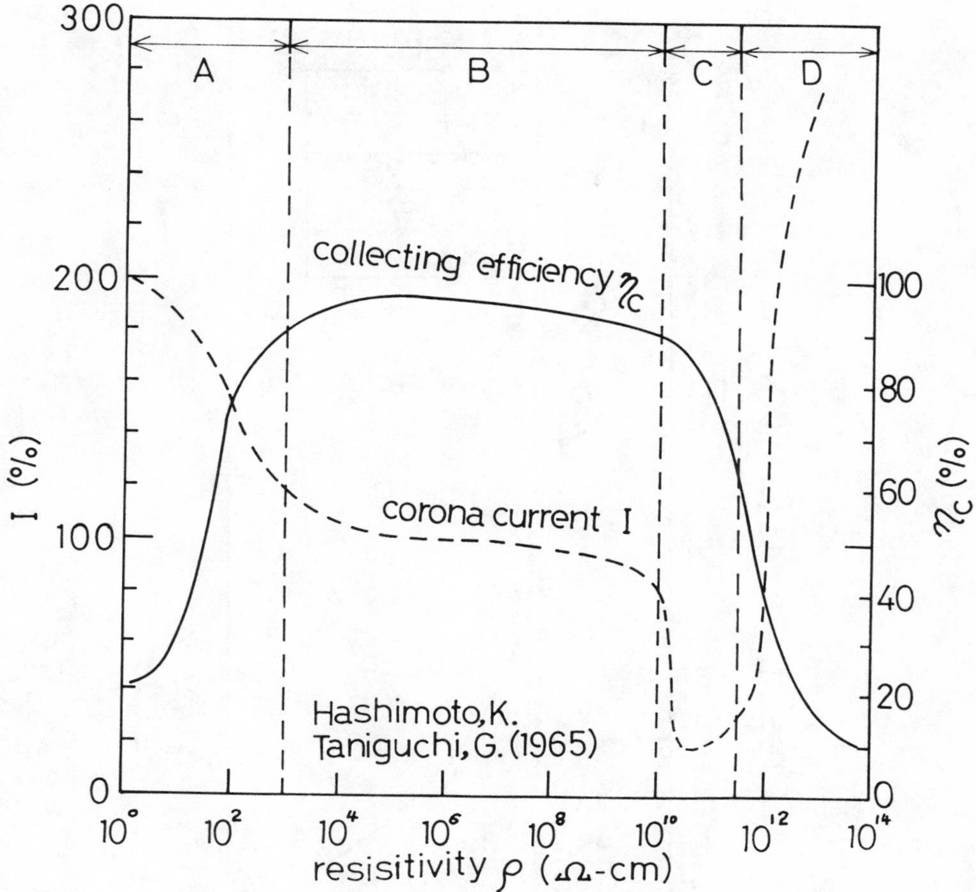

FIGURE 26. Relationship between the collection efficiency and the resistivity of the materials.

to the decreasing of the void inside the dust layer. But from the practical point of view, the fine solid particles or fumes have not only various particle size Xp, but also various chemical components. Therefore, decreasing the particle size Xp, the resistivity is not always decreased. This typical and practical example is shown in Figure 29. This figure shows the relationship between the resistivity ρ of boiler fly-ash and the temperature T which is a function of the particle diameter Xp.

XIII. THEORETICAL ESTIMATION OF THE COLLECTION EFFICIENCY

The fundamental idea of the theoretical estimation of the collection efficiency η_c which was derived by Deutsch (1922)[27] is described in this section.

As shown in Figure 30, the feed concentration of dust at the entrance is Co(g/m^3) and the migration velocity of the particle diameter Xp is $\overline{U}$p(m/s). The area of the collecting electrode plate is H·L (m^2). Assuming that the dust concentration C(g/m^3) and the gas velocity Vo(m/s) at the distance X(m) distribute homogeneously between the collecting electrodes, the collecting mass dM(g) on the small distance dX, and the time dt = dX/$\overline{V}$o for the flow direction can be written

$$dM = 2H \cdot dX \cdot C \cdot \overline{U}p \cdot dt = 2H \cdot C \cdot \overline{U}p \cdot \frac{(dX)^2}{\overline{V}o} \tag{24}$$

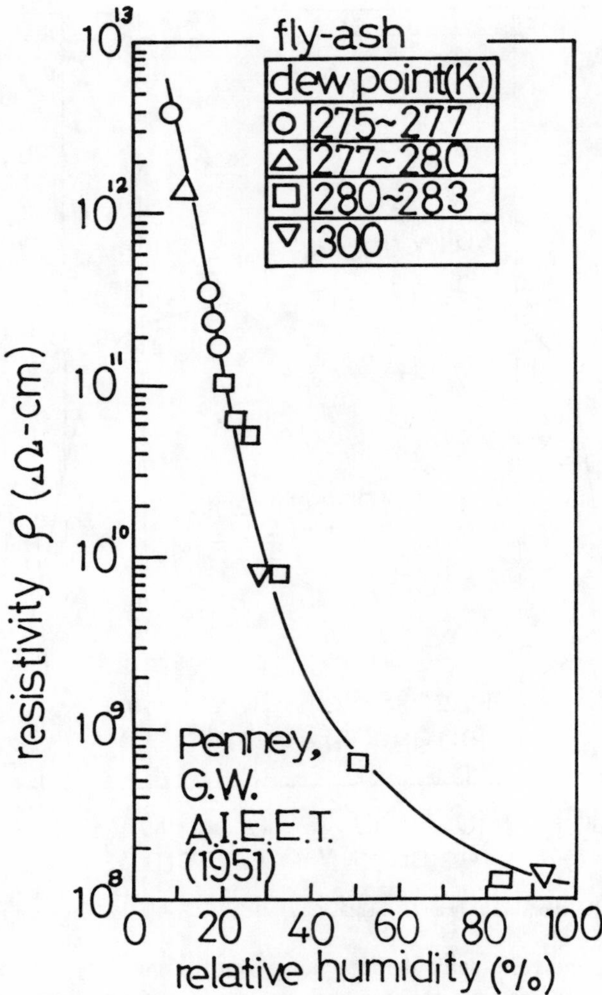

FIGURE 27. Resistivity of fly-ash for humidity.

On the other hand, denoting that the cross-sectional area of the electrostatic precipitator is B·H, then we can obtain the mass balance equation

$$dM = -dC \cdot B \cdot H \cdot dX \tag{25}$$

Substituting Equation 24 into Equation 25, we can obtain the following equation

$$\frac{dC}{C} = -\frac{2 \cdot \bar{U}p \cdot dX}{B \cdot \bar{V}o} \tag{26}$$

Consequently, integrating Equation 26 for the boundary conditions as C = Co at X = 0 and $C = C_L$ at X = L, we can obtain the concentration equation

$$\frac{C_L}{Co} = \exp\left(-\frac{2 \cdot \bar{U}p \cdot L}{B \cdot \bar{V}o}\right) \tag{27}$$

Therefore the theoretical collection efficiency η_c can be obtained:

$$\eta_c = \frac{Co - C_L}{Co} = 1 - \exp\left(-\frac{2 \cdot \bar{U}p \cdot L}{B \cdot \bar{V}o}\right) \tag{28}$$

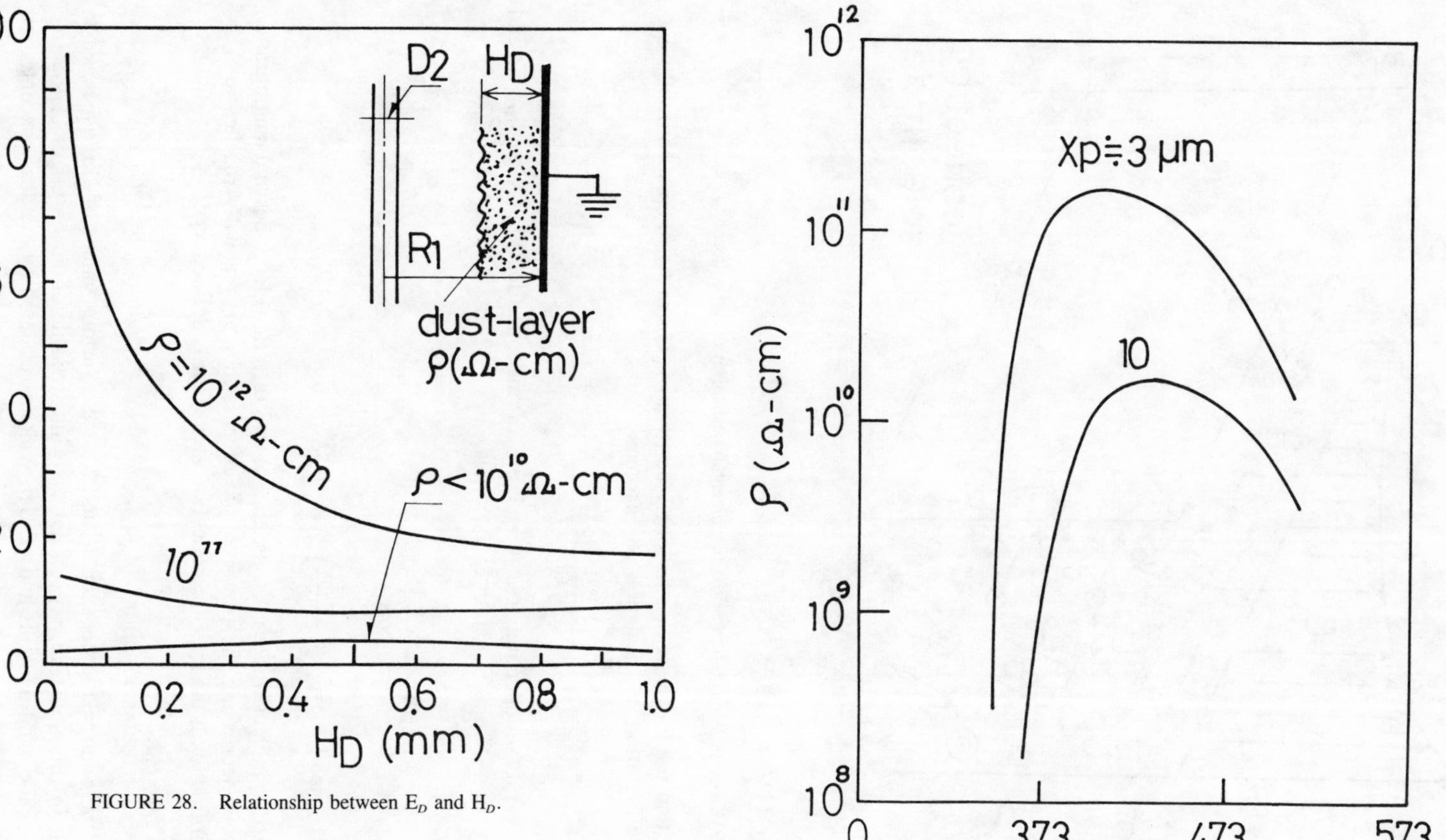

FIGURE 28. Relationship between E_D and H_D.

FIGURE 29. Relationship between ρ and T for fly-ash of a boiler by Hashimoto and Taniguchi (1965).

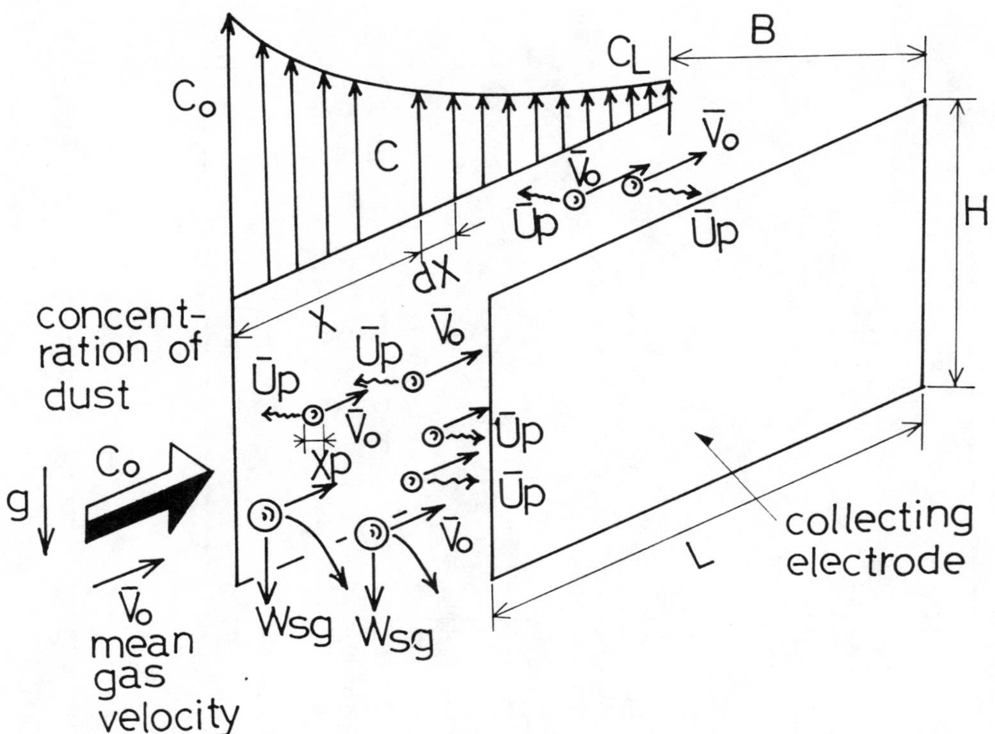

FIGURE 30. Illustration of the fundamental mechanism of the collection of dust.

Here denoting that the gas flow rate for one lane (duct) is $Qo(m^3/s)$, then substituting $\overline{V}o = Qo/B{\cdot}H$ into Equation 28, Equation 28 becomes

$$\eta_c = 1 - \exp\left(-\frac{2{\cdot}H{\cdot}L{\cdot}\overline{U}p}{Qo}\right) \tag{29}$$

Substituting the total collecting area $A = 2{\cdot}B{\cdot}H$ of one lane (duct) into Equation 27 and 29, finally Equation 27 and Equation 29 becomes

$$\frac{C_L}{Co} = \exp\left(-\frac{A{\cdot}\overline{U}p}{Qo}\right) \tag{30}$$

$$\eta_c = 1 - \exp\left(-\frac{A{\cdot}\overline{U}p}{Qo}\right) \tag{31}$$

From this equation, increasing area A (one lane) of the collecting electrodes and the migration velocity $\overline{U}p$, but decreasing the flow rate Qo of gas, then η_c gradually reaches to $\eta_c \doteqdot$ 100%. Here, introducing the specific collection area f(s/m), defined as

$$f(s/m) = \frac{A}{Qo} \tag{32}$$

we can obtain the relationship between η_c and f for the various values of Up, as shown in Figure 31. But in this equation, we do not take into consideration the effect of the free sedimentation of the solid particles and of $\overline{U}p$, which is effected by the dust concentration in the electrostatic precipitator.

On the other hand, Masuda (1966) considered the general theory of statistical design and its application to the electrostatic precipitator.[28] However, in this context, we do not need to refer to this domain.

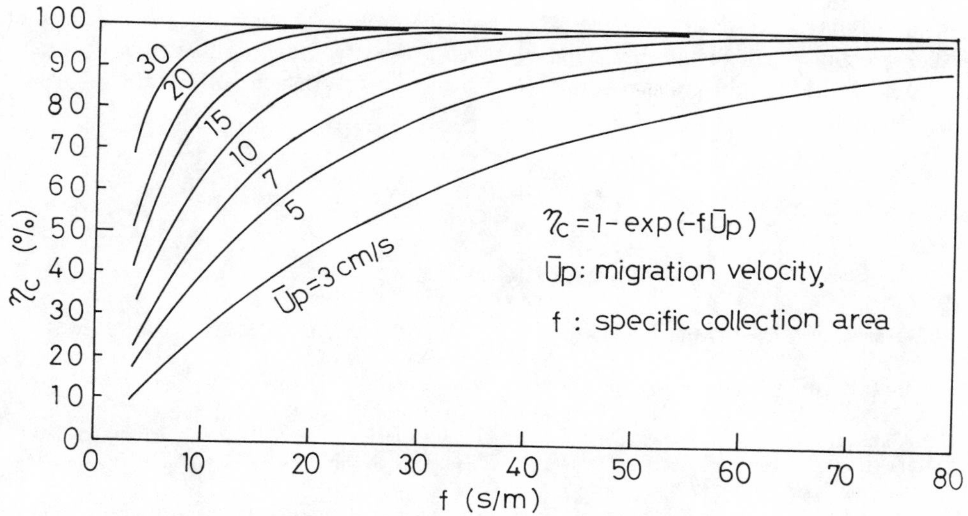

FIGURE 31. Theoretical relationship between f and η_c.

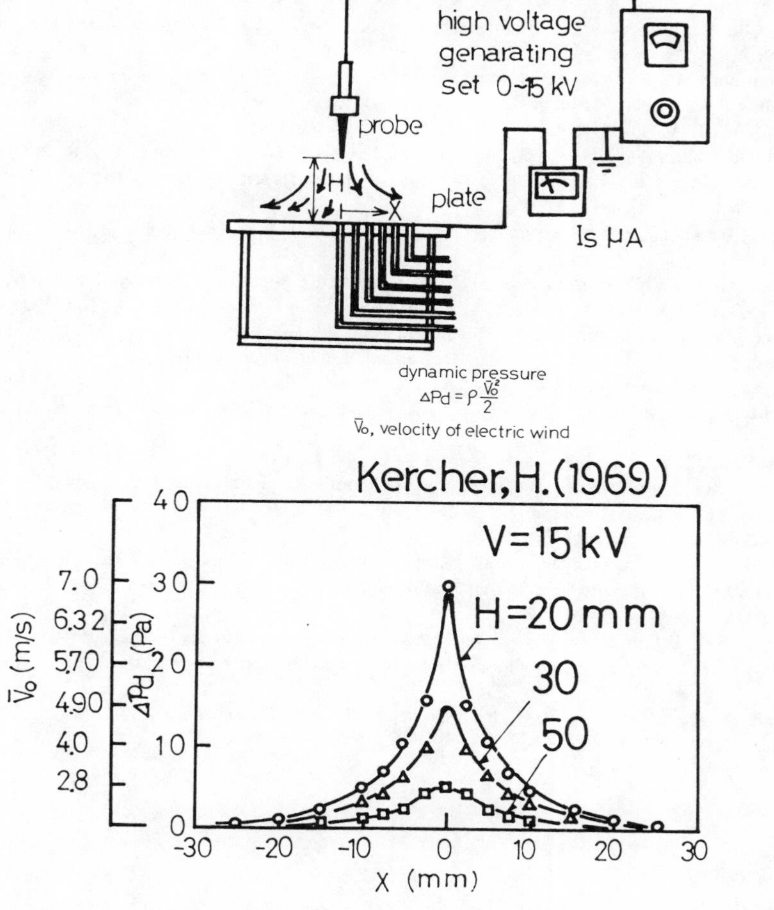

FIGURE 32. Experimental results of electric wind.

In addition to this, Figure 32 shows the experimental results of the electric wind by Kercher (1969).[29] From this experiment, the fluid velocity Vo of electric wind depends on the electric field. Also when the distance between the probe and the plate is H = 20 mm, then the velocity Vo of the electric wind becomes to Vo = 7 m/s at the center position. Therefore, electric wind may strongly affect the motion of charged solid particles in the electric precipitator.

REFERENCES

1. **Marchello, J. M. and Kelly, J. J.,** *Gas Cleaning For Air Quality Control,* Marcel Dekker, New York, 1975.
2. **Bethea, R. M.,** *Air Pollution Control Technology,* Van Nostrand, New York, 1979.
3. **Parker, A.,** *Industrial Air Pollution Handbook,* McGraw-Hill, New York, 1978.
4. **Fukuda, S.,** Electrical precipitation and its application (in Japanese), *Rep. I.E.E. Jpn.,* January, 1, 1930.
5. **Katz, J.,** Maintenance and operation problems of electrostatic precipitators, *J. Air Pollut. Control Assoc.,* 28(9), 868, 1978.
6. **Hall, H. J.,** Solving problems in the electrical energization of electrostatic precipitators, *J. Air Pollut. Control Assoc.,* 28(9), 870, 1978.
7. **Greco, J., Ezzell, R. O., and Lytle, J. H.,** An integrated program to achieve maximum performance of electrostatic precipitators, *J. Air Pollut. Control Assoc.,* 28(9), 877, 1978.
8. **Loeb, L. B.,** *Fundamental Processes of Electrical Discharge in Gases,* John Wiley & Sons, New York, 1939.
9. **Matzumoto, T.,** *Electrostatic Precipitators* (in Japanese), Nikankogyo-Shinbun-Sha, Tokyo, 1980.
10. **Loeb, L. B.,** *Kinetic Theory of Gases,* McGraw-Hill, New York, 1927.
11. **Millikan, R. A.,** in *Phys. Rev.,* 22(2), 1, 1923.
12. **Seidenki-Gakai,** *Handbook of Electrostatics* (in Japanese), OHM Publishing, Tokyo, 1981.
13. **Fukuda, S.,** On the charging-up of particles in electric precipitators (in Japanese), *Rep. I.E.E. Jpn.,* September, 943, 1928.
14. **White, H. J.,** The role of corona discharge in the electrical precipitation process, *Electr. Eng. (Am. Inst. Electr. Eng.),* June, 67, 1952.
15. **Matzumoto, T.,** *Electrostatic Precipitators* (in Japanese), Nikankogyo-Shinbun-Sha, Tokyo, 1980.
16. **Bahtaeb, SH. A. and Grinman, I. G.,** *Koronnorazriadnye Privory,* Nauka, Russia, 1975.
17. **Fasel, C. S. and Parsons, S. R.,** The current-voltage relation in the corona, *Phys. Rev.,* 23, 598, 1924.
18. **Mayr, O.,** Raumladungs Problem der Hochspannungstechnik, *Arch. Elektrotech. (Berlin),* 18, 270, 1927.
19. **Bright, A. W.,** Fundamentals of dust precipitation in an electrostatic field, *Filtr. Sep.,* June, 284, 1979.
20. **White, H. J.,** Particle charging in electrostatic precipitation, *Am. Inst. Electr. Eng. Trans.,* 70, 1186, 1951.
21. **Takahashi, K.,** *Fundamental Aerosol Engineering* (in Japanese), Yokendo, Tokyo, 1972.
22. **Sproull, W. T. and Nakada, Y.,** Operation of Cottrell precipitators, *Ind. Eng. Chem.,* 43(6), 1350, 1951.
23. **Townsend, J. S.,** in *Philos. Mag.,* 28, 83, 1914.
24. **Tihodeev, N. N.,** *Zh. Tehn. Fiz.,* 25, 8, 1955.
24. **Bright, A. W.,** Fundamentals of dust precipitation in an electrostatic field, *Filtr. Sep.,* June, 284, 1979.
26. **Penney, G. W.,** Electrostatic precipitation of high resistivity dust, *Am. Inst. Electr. Eng. Trans.,* 70(9), 1192, 1951.
27. **Koglin, W.,** Die Lastabhängigkeit von Elektrofilteranlagen, *Staub,* 21(5), 212, 1961.
28. **Masuda, S.,** Statistische Betrachtungen über den Abscheidegrad des Elektrofilters, *Staub,* 26(11), 459, 1966.
29. **Kercher, H.,** Elektischer Wind, Rücksprüken und Staubwiderstand als Einflußgrößen im Elektrofilter, *Staub,* 29(8), 314, 1969.

Chapter 5

WET SCRUBBER

I. INTRODUCTION

The wet scrubber which washes fine dust off by using water droplets is a wet-type of particulate collector. The collection mechanisms are approximately the same as in filtration, i.e., impaction, interception, and diffusion.[1,2]

A. Impaction

From the theoretical point of view according to Kleinschmidt[3] (1939), the collection efficiency η_c could be written

$$\eta_c = 1 - \exp\left(-\frac{3 \cdot \eta_t (Ql/Qg) \cdot X}{2 \cdot Dp}\right) \tag{1}$$

where η_t is target efficiency, X is distance, Ql/Qg is ratio of flow rate of liquid to gas, and Dp is droplet diameter. When the droplet diameter Dp($Dp \geqq 200$ Xp) is sufficiently larger than particle diameter Xp, the target efficiency η_t becomes $\eta_t \rightarrow 1$.

B. Diffusion

The fine solid particles contact with the droplet surface by Brownian motion. This effect becomes remarkable for the particle size Xp smaller than 0.5 μm. At the same time the turbulent diffusion plays an important influence in the collection of particles.

C. Humidity

The coagulation of the particles to each other is promoted by changing the electrification on the particles while increasing the humidity. Consequently it becomes easy to separate the coagulated particles.

D. Condensation

When the temperature of the fume decreases under the dew point by the injection of cold water, the particle size becomes larger with the condensation on the particle. This phenomenon becomes remarkable in the case of high temperature before injection and in the case of the dust concentration lower than 2 g/m^3.

E. Wet

It may be possible to increase the collection efficiency by giving an active surface agent to the water to reduce its surface tension.

Wet scrubbers can separate easily until about the particle size of Xp = 1 μm on the pressure drop Δp_c = 2 kPa. The construction of this scrubber is simple and inexpensive, however, it is very expensive for drain treatment. Also, it is very expensive to operate for the purpose of separating sub-micron particles (Xp = 0.1 to 0.4 μm) by applying the types of venturi scrubbers using a high pressure drop (Δp_c = 3 to 20 kPa). Here Figure 1 shows the general characteristics of the five types of wet scrubbers which were investigated by Wicke and Krebs in 1971.[4]

II. CYCLONIC-SPRAY SCRUBBER

Pease-Anthony's gas scrubber (cyclonic-spray scrubber), as shown in Figure 2, was developed in 1930 for boilers by M. D. Engle.[5,6] Two cyclonic-scrubbers for flue gases and

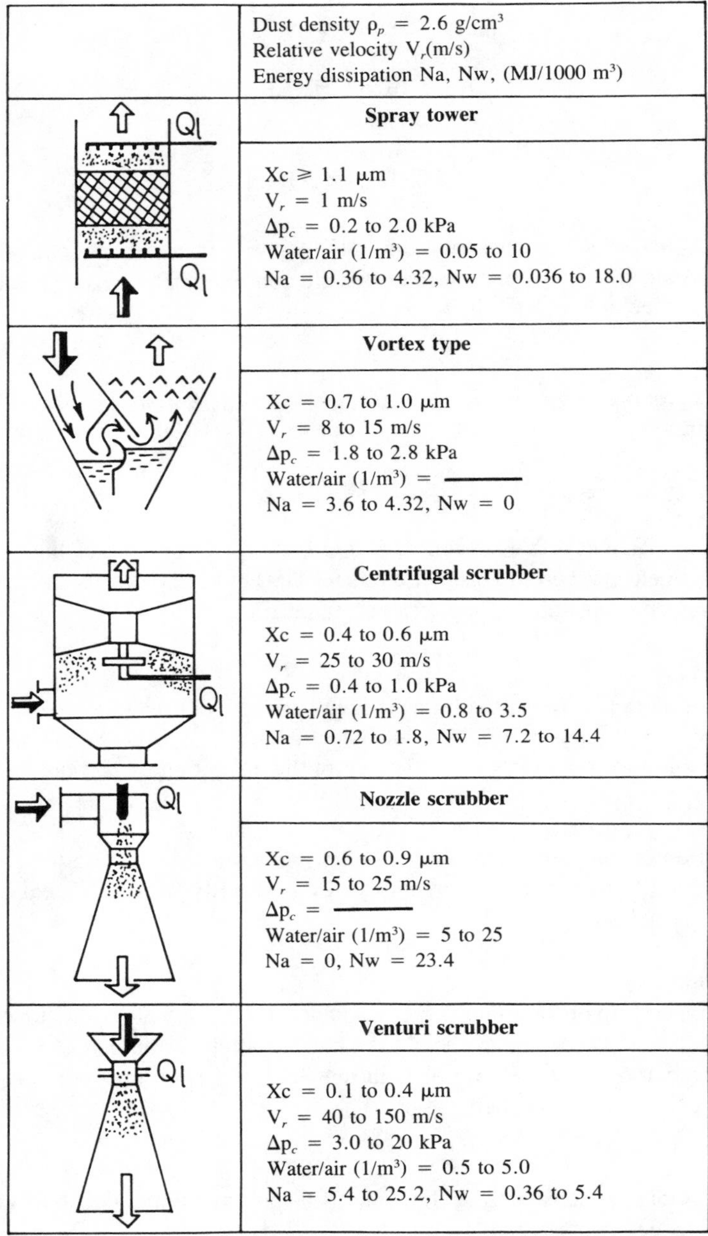

	Dust density ρ_p = 2.6 g/cm³ Relative velocity V_r(m/s) Energy dissipation Na, Nw, (MJ/1000 m³)
	Spray tower
	Xc ≥ 1.1 μm V_r = 1 m/s Δp_c = 0.2 to 2.0 kPa Water/air (1/m³) = 0.05 to 10 Na = 0.36 to 4.32, Nw = 0.036 to 18.0
	Vortex type
	Xc = 0.7 to 1.0 μm V_r = 8 to 15 m/s Δp_c = 1.8 to 2.8 kPa Water/air (1/m³) = —— Na = 3.6 to 4.32, Nw = 0
	Centrifugal scrubber
	Xc = 0.4 to 0.6 μm V_r = 25 to 30 m/s Δp_c = 0.4 to 1.0 kPa Water/air (1/m³) = 0.8 to 3.5 Na = 0.72 to 1.8, Nw = 7.2 to 14.4
	Nozzle scrubber
	Xc = 0.6 to 0.9 μm V_r = 15 to 25 m/s Δp_c = —— Water/air (1/m³) = 5 to 25 Na = 0, Nw = 23.4
	Venturi scrubber
	Xc = 0.1 to 0.4 μm V_r = 40 to 150 m/s Δp_c = 3.0 to 20 kPa Water/air (1/m³) = 0.5 to 5.0 Na = 5.4 to 25.2, Nw = 0.36 to 5.4

FIGURE 1. General characteristics of five types of wet scrubbers.

fly-ash removal from boilers were installed in 1930, each rated at 34.2 m³/s at 541 K. The mean axial velocity Vz of gas in scrubber is about Vz = 0.5 to 2.5 m/s. At the time of Engle's tests, these units had about 90 centrifugal-type nozzles of 4.76 mm orifice, operating at about 0.38 MPa, and showed full-load efficiency η_c = 80 to 82%. Disregarding requirements for the saturation of gases, 180 efficient nozzles of 1.6 mm orifice at the same pressure would pass about 25% of the amount of water, but the efficiency would rise to about η_c = 88%. If 270 similar nozzles were used, with 37.5% of the water, the efficiency η_c would be about 93%.

Figure 3 shows the size distribution of water droplets from a centrifugal nozzle of diameter

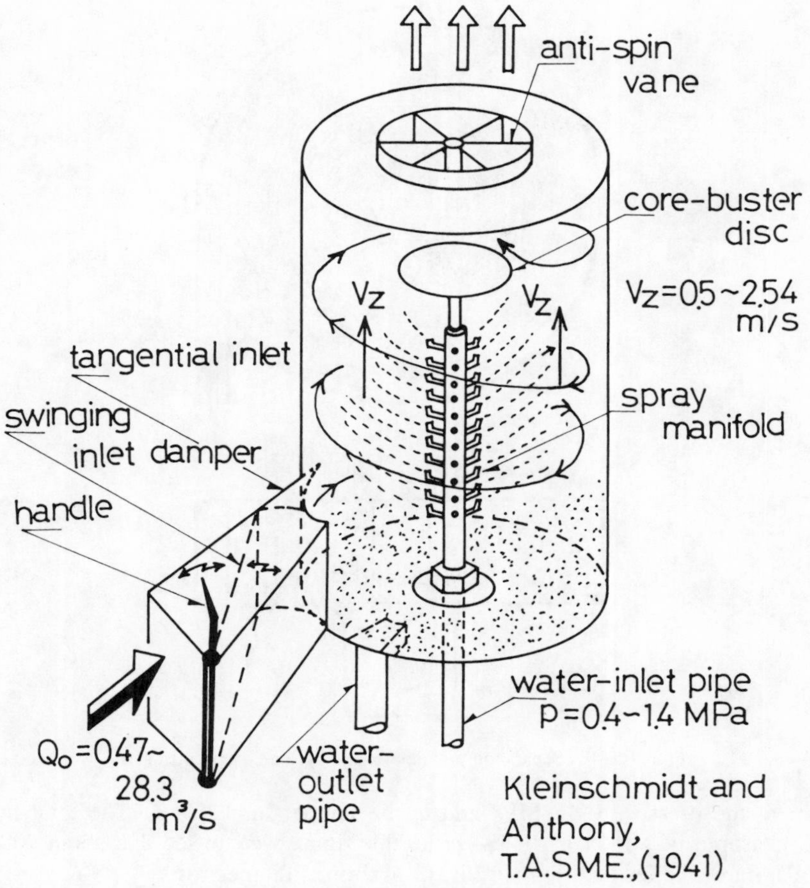

FIGURE 2. Cyclonic-spray scrubber.

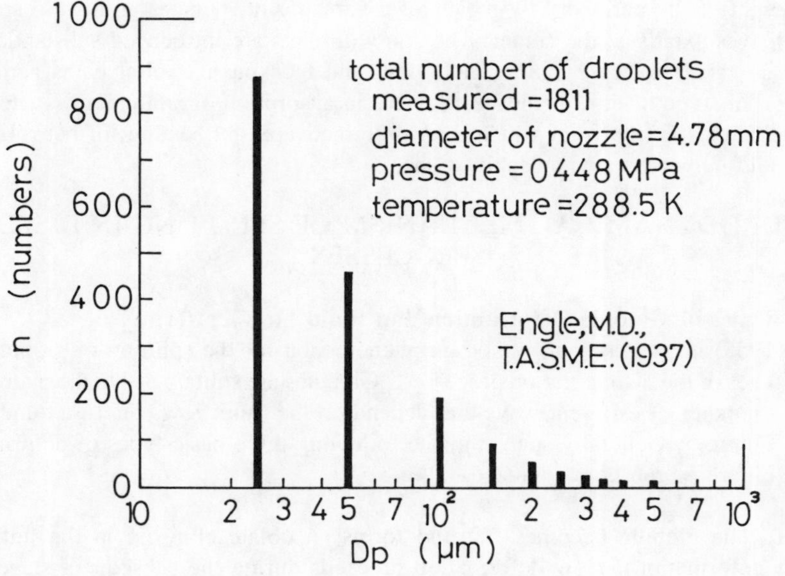

FIGURE 3. Size distribution of water droplets by centrifugal nozzle.

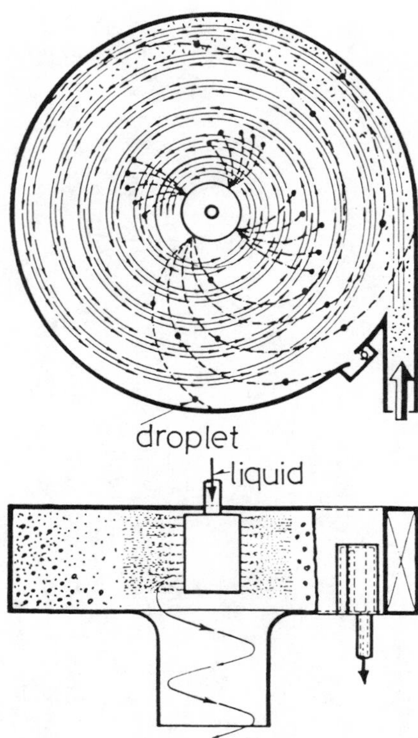

FIGURE 4. Separation principle of the cyclonic-spray scrubber.

4.78 mm at the pressure 0.45 MPa and at the temperature 289 K. The total number of droplets measured was 1811 particles. From this figure we can see that about 900 droplets (about 50%) are smaller than Dp = 25 μm in droplet diameter.

Figure 4 shows the cyclonic-spray scrubber principle which obtains the true countercurrent contact between gas and liquid with several obvious advantages. The dust-laden gas enters the full height of the periphery by means of a narrow slot, pursues an inward spiral path, and then leaves axially at the center. The spray droplets are introduced substantially along the axis, are taken up the spinning gas body, and then pursue spiral paths across to the periphery. This type of unit would be effective in absorption of ammonia in water, of H_2S in weak alkaline solution from which it is to be recovered by heating, of benzol vapors in wash oil, and numerous similar applications.

III. HYDRODYNAMICAL MECHANISM OF SPLITTING IN DISPERSION PROCESSES

A. Basic Types of Globule Deformation and Fluid Flow Pattern

Hinze (1955) investigated an important phenomenon for the splitting of globules during the final stages of the disintegration processes.[7] Globules are split up due to the hydrodynamic forces in a number of different ways that depend on the fluid flow pattern around them, as shown in Figure 5. Generally speaking, the following three basic types of deformation, as shown in Figure 5, can be recognized.

Type 1. The globule becomes flat and forms an oblate ellipsoid in the initial stages (lenticular deformation). How deformation proceeds during the subsequent stages leading to breakup seems to depend on the magnitude of the external forces causing the deformation.

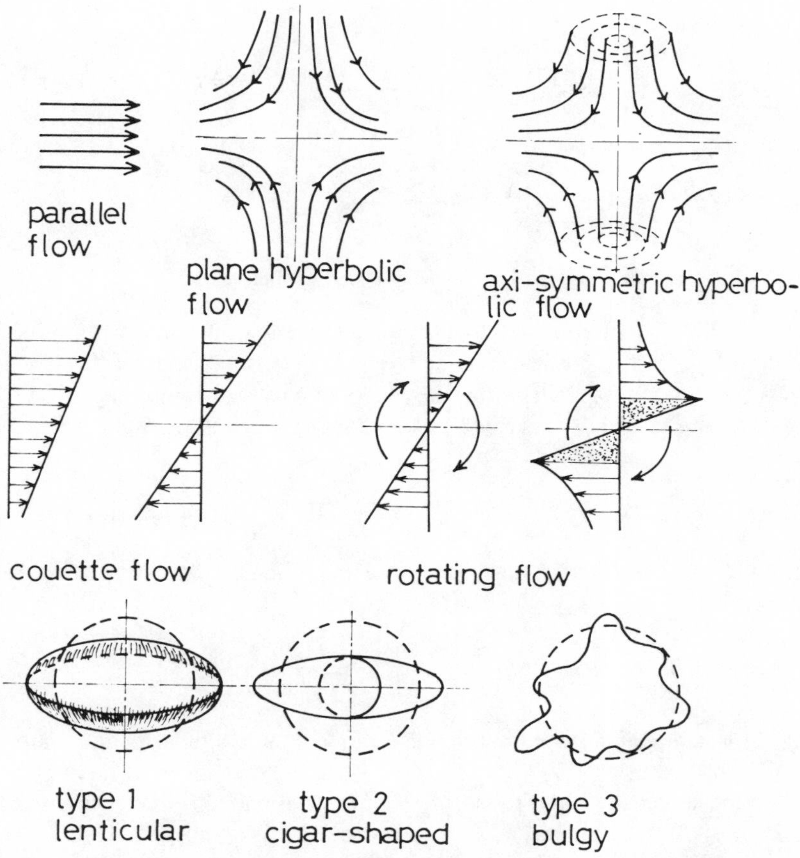

FIGURE 5. Basic types of globule deformation and fluid flow pattern.

Type 2. The globule becomes more and more elongated and forms a prolate ellipsoid in the initial stages, and then ultimately a long cylindrical thread which is broken up into many small droplets (cigar-shaped deformation).

Type 3. The surface of the globule is deformed locally, bulges and protuberances occur, and parts of the globule become separated (bulgy deformation).

Therefore, it may be expected that the deformation of Type 1 will occur if the globule is subjected to the dynamic pressures, or viscous stresses, produced by the parallel fluid flow (see Figure 5), the axisymmetric hyperbolic fluid flow, and the rotating fluid flow. The deformation of Type 2 (cigar-shaped deformation) can be caused by (refer to Figure 5) the plane hyperbolic fluid flow, and the Couette flow, and still more the bulgy deformation of Type 3 is brought about only by the dynamic pressures occurring in the irregular flow pattern.

Breakup will occur if there is a sufficient degree of deformation, that is, if the region of the flow pattern causing a specific deformation is sufficiently large to contain the deformed globule and if the flow pattern persists long enough.

B. Forces Splitting Up a Globule

An isolated globule will be considered on a surface on which external forces act to cause its deformation. Here, τ (N/m^2) is a force per unit surface area. This force will vary along the surface and will be a function of time. It may be a viscous stress on a dynamic pressure set up in the surrounding continuous fluid phase. Furthermore, the interfacial tension σ (N/m) will give rise to a surface force that will in general counteract the deformation.

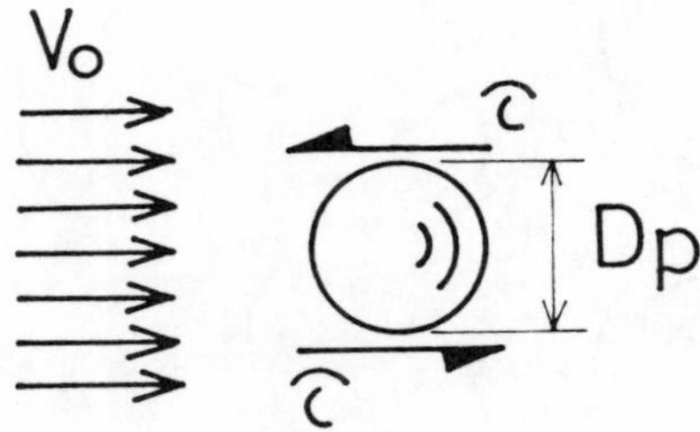

FIGURE 6. Illustration of a force on a globule.

Now if Dp is the diameter of the globule, the surface-tension force will be of the order of magnitude σ/Dp. In the first instance the dynamic pressure will be of the same order of magnitude as τ which corresponds to the flow velocity in the order of magnitude $Vo \doteqdot (\tau/\rho_p)^{1/2}$. Therefore, the viscous stresses are in the order of magnitude, as shown in Figure 6

$$\frac{\eta_p}{Dp}\sqrt{\frac{\tau}{\rho_p}} \quad (2)$$

where η_p and ρ_p are the viscosity and density of the globule. Thus these three forces per unit area

$$\tau,\ \sigma/Dp \text{ and } \frac{\eta_p}{Dp}\sqrt{\frac{\tau}{\rho_p}}$$

control the deformation and breakup of the globule, and three dimensionless groups may be formed.

For one of the dimensionless groups, the condition (generalized Weber group) We = τ·Dp/σ is chosen. In general, the Weber group usually refers to $\rho Vo^2 \cdot Dp/\tau$, where ρVo^2 means the dynamic pressure of fluid flow. For the other dimensionless group either

$$\frac{\eta_p}{\sigma}\cdot\sqrt{\frac{\tau}{\rho_p}} \text{ or } \frac{\tau\cdot Dp}{\eta_p}\cdot\sqrt{\frac{\rho_p}{\tau}} = \frac{\rho_p\cdot Dp}{\eta_p}\cdot\sqrt{\frac{\tau}{\rho_p}}$$

may be chosen, the latter being the Reynolds group. For the last dimensionless group, a viscosity group

$$Vi = \frac{\eta_p}{\sqrt{\rho_p\cdot\sigma\cdot Dp}}$$

may be chosen because it accounts for the effect of the viscosity of the fluid in the globule.

Then the experiments have shown that the glouble breakup mechanism is quite simple when We is equal to, or slightly higher than, a critical value We.crit at which the breakup of the globule occurs. The more We exceeds its critical value, the more complicated the mechanism. For We ≫ We.crit, this mechanism becomes very complex and the disintegration process is more or less chaotic.

C. Dynamic Model of the Breakup of Drops in Turbulent Flow

The dispersion of one liquid into another as the result of keeping the latter in violent turbulent motion is restricted here to the case of emulsification under noncoalescing conditions (for instance, by taking low concentration of the dispersed phase so that the chance of coalescing is very slight).

Here, the kinetic energy of a turbulent fluctuation increases with increasing wave length. Thus velocity differences due to the fluctuations with a wave length equal to 2·Dp will produce a higher dynamic pressure than those due to fluctuations with a shorter wave length. If these fluctuations are assumed to be responsible for the breakup of drops for We.crit may be placed as

$$\text{We.crit} = \frac{\rho \cdot \overline{v^2} \cdot \text{Dp.max}}{\sigma} \tag{3}$$

where $\overline{v^2}$ is the average value across the whole flow field of the squares of velocity differences over a distance equal to Dp.max. To relate this average kinetic energy to this distance, Hinze considered the simplest case, namely an isotropic homogeneous turbulence. For this case of turbulence, the main contribution to the kinetic energy is made by the fluctuations in the region of wave lengths where Kolmogoroff's energy distribution law is valid. In this region, the turbulence pattern is solely determined by the energy input $\epsilon(\text{cm}^2/\text{s}^3)$ per unit mass and unit time. It can be shown that for this region

$$\overline{v^2} = C_1 \, (\epsilon \cdot \text{Dp})^{2/3} \tag{4}$$

where $C_1 \doteqdot 2.0$ according to Batchelor (1951). If for the moment one assumes that Vi $\ll$ 1, then one obtains from Equation 3

$$\text{We.crit} = \frac{\rho \cdot \text{Dp.crit}}{\sigma} \cdot C_1 \cdot (\epsilon \cdot \text{Dp.max})^{2/3} = \text{Const.} \tag{5}$$

or

$$\text{Dp.max} \cdot \left(\frac{\rho}{\sigma}\right)^{3/5} \cdot \epsilon^{2/5} = C \tag{6}$$

From this equation we can see that only the quantities ρ, σ, and ϵ determine the size Dp.max of the largest drops.

Then if in the emulsifying apparatus, the flow field is not too inhomogeneous, the powerful diffusive action of turbulence causes the average size of the largest drop in the whole field to correspond to the average energy input across this field. With these assumptions, Equation 6 has been applied to the results obtained by Clay[8] (1940) with his model arrangement consisting of two coaxial cylinders, the inner one of which is rotated. Hinze obtained the value C of Equation 6 as

$$\text{Dp.max} \left(\frac{\rho}{\sigma}\right)^{3/5} \cdot \epsilon^{2/5} = 0.725 \tag{7}$$

with a standard deviation of 0.315. Finally he obtained a relationship between $\rho \cdot \sigma \cdot \text{Dp.max}/\eta^2$ and $\eta^5 \cdot \epsilon/\rho \cdot \sigma^4$ for $10^{-13} \leqq \eta^5 \cdot \epsilon/\rho \cdot \sigma^4 \leqq 10^{-4}$ as follows:

$$\frac{\rho \cdot \sigma \cdot \text{Dp.max}}{\eta^2} = 0.725 \left(\frac{\eta^5 \cdot \epsilon}{\rho \cdot \sigma^4}\right)^{-2/5} \tag{8}$$

Equation 8 is shown in Figure 7.

IV. SMALL PARTICLE COLLECTION BY SUPPORTED LIQUID DROPS

The phenomenon of particle collection by objects of different shapes has been the subject of many experimental and theoretical studies.[9] Three different collection mechanisms — inertia, diffusion, and interception — have been postulated and proved. This study was limited to the collection by inertia mechanisms only, thus very high air velocity Vo = 9 to 17 m/s and very low Xp/Dl ratios were employed, where Xp and Dl are the solid particle and droplet diameters, respectively.

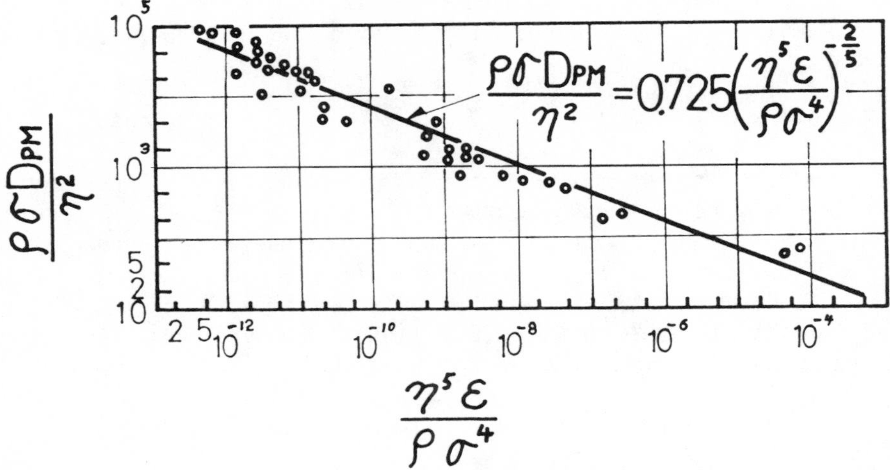

FIGURE 7. Comparison of the experimental results with Equation 8.

Figure 8 shows the experimental results of the collection efficiency by glycerol drops for polystyrene particles of diameters Xp = 0.814, 1.171, 1.80, and 2.85 μm. In this figure, the experimental result of Walton and Woolcock[10] from 1960 is shown by a dotted line. Here symbol Ko means a dimensionless number defined as

$$Ko = \frac{\lambda}{\lambda s} \cdot \frac{2 \cdot \rho_p \cdot Vo \cdot Xp^2}{18 \cdot \eta \cdot Dl} \cdot Cn \tag{9}$$

where Cn is Cunningham slip correction factor and $\lambda/\lambda s$ is the ratio of the distance for which a particle will travel if projected into still air with velocity Vo to the range it would have if the drag force were that of Stokes drag.

V. PHYSICAL MECHANISMS OF PARTICLE COLLECTION

A. Sprays

Particle collection by liquid spray drops are accomplished by several mechanisms, but inertia is the force most effective in scrubbers. The author will describe the mathematical model from a paper of Fonda and Herne (1960).[11,12]

B. Cross-Flow

In the case of the cross-flow as shown in Figure 9, the spray will fall a distance H with a velocity Wlg and will be evenly distributed over a length Z. The volume fraction of drops at any point q (drop holdup) is given by a spray material balance for a differential section of the scrubber as

$$u_l \cdot H \cdot B \cdot q - u_l \cdot H \cdot B \cdot \left(q + \frac{dq}{dz} \cdot dz\right) + \frac{Qw \cdot B \cdot dz}{B \cdot Z} - Wlg \cdot q \cdot B \cdot dz = 0 \tag{10}$$

where Qw is the volumetric flow rate of liquid and u_l is the velocity for the horizontal direction. Here, assuming that dq/dz = 0, then from Equation 10 we can obtain

$$q = \frac{Qw}{Wlg \cdot B \cdot Z} \tag{11}$$

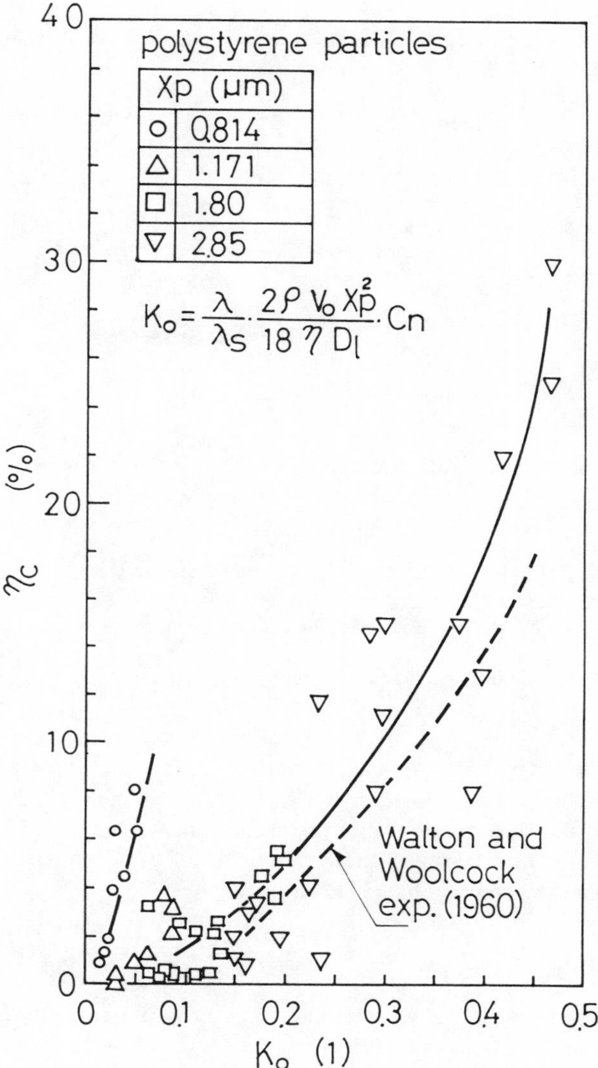

FIGURE 8. Experimental results of the collection efficiency by the glycerol drops for the polystyrene particles.

On the other hand, a material balance of dust over the same volume element can be written

$$up \cdot H \cdot B \cdot C - up \cdot H \cdot B \cdot \left(c + \frac{dC}{dz} \cdot dz\right) - Wlg \cdot H \cdot q \cdot E \cdot \left(\frac{3}{4r}\right) \cdot C \cdot B \cdot dz = 0 \tag{12}$$

where 3/4r is the ratio of drop cross-sectional area to drop volume

$$\frac{3}{4r} = \frac{\pi \cdot Dl}{\frac{4}{3} \cdot \pi \cdot \left(\frac{Dl}{2}\right)^3 \cdot 4}$$

therefore H·B·dz·(3/4r)·q means the total cross-sectional area of drops, u*p* is the horizontal velocity, E is the particle collection efficiency of drop for particles of one size, and C is

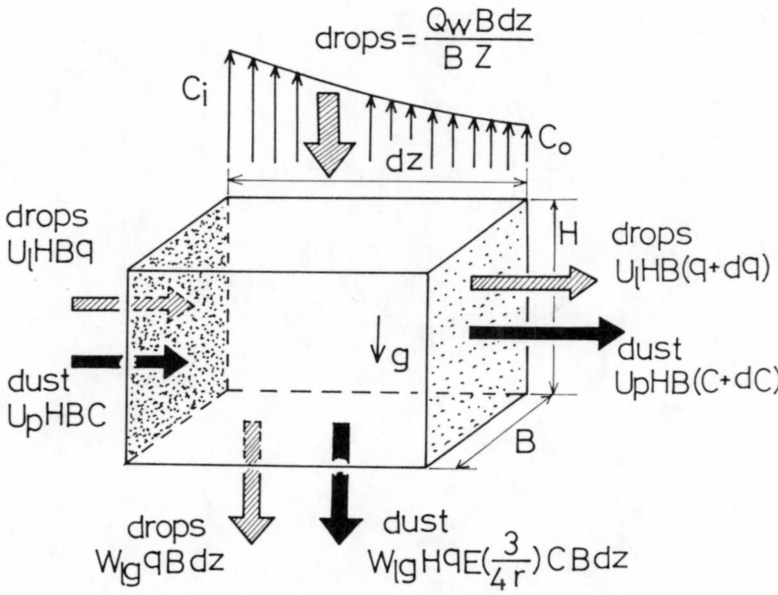

FIGURE 9. Mathematical model for the cross-flow.

the dust concentration. Consequently, the differential equation of the dust concentration can be written

$$\frac{dC}{C} = -\frac{3 \cdot E \cdot q \cdot Wlg}{4 \cdot r \cdot up} \cdot dz \tag{13}$$

Then assuming that the dust concentration across any plane perpendicular to the flow is uniform and Ci and Co are the dust concentrations at inlet and outlet, we can obtain the equation of dust concentration combined Equation 13 as

$$\ln \frac{Co}{Ci} = -\frac{3 \cdot E \cdot q \cdot Wlg \cdot Z}{4 \cdot r \cdot up} = -\frac{3 \cdot E \cdot H \cdot Qw}{4 \cdot r \cdot Qg} \tag{14}$$

where Qg = *up* H·B means the gas flow rate. Therefore the total collection efficiency η_c can be calculated as:

$$\eta_c = \frac{Ci - Co}{Ci} = 1 - \exp\left(-\frac{3 \cdot E \cdot H \cdot Qw}{4 \cdot r \cdot Qg}\right) \tag{15}$$

We will find from this equation that the collection efficiency η_c increases with the height of the drop fall of spray-chamber and with increasing droplet flow rate and also with decreasing flow rate Qg of the dust-laden gas flow.

C. Vertical Countercurrent Flow

In the vertical countercurrent flow scrubber[13] of height H and cross-sectional area A, as shown in Figure 10, the drops of spray and dust balance can be written

$$q = \frac{Qw}{(Wlg - Vg) \cdot A} = \frac{Qw}{v_r \cdot A} \tag{16}$$

where v_r is the relative velocity between Wlg and v_g,

$$\frac{dC}{C} = -\frac{3 \cdot Wlg \cdot q \cdot E \cdot dZ}{4 \cdot r \cdot v_g} \tag{17}$$

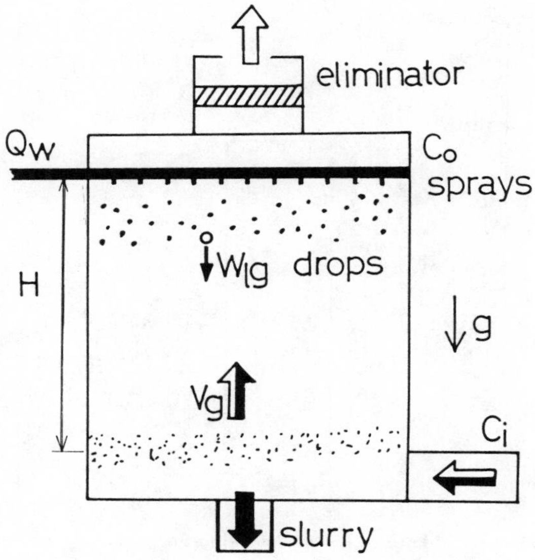

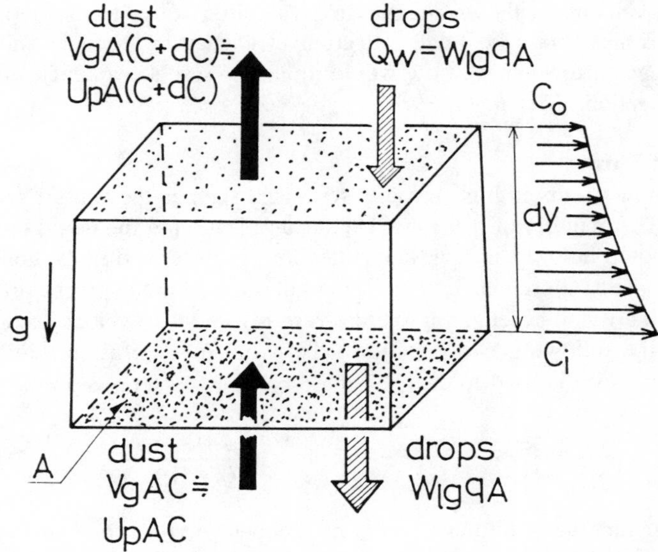

FIGURE 10. Mathematical model for the vertical countercurrent flow.

This equation can be integrated as

$$\ln \frac{Co}{Ci} = -\frac{3 \cdot Wlg \cdot q \cdot E \cdot H}{4 \cdot r \cdot v_g} \tag{18}$$

and then Equation 18 can be transformed by applying Equation 16 as

$$\ln \frac{Co}{Ci} = -\frac{3 \cdot Wlg \cdot E \cdot H}{4 \cdot r \cdot v_g} \cdot \frac{Qw}{v_r \cdot A} = -\frac{3 \cdot Wlg \cdot E \cdot H \cdot Qw}{4 \cdot r \cdot v_r \cdot Qq} \tag{19}$$

Therefore the total collection efficiency η_c can be obtained as

$$\eta_c = \frac{Ci - Co}{Co} = 1 - \frac{Co}{Ci} = 1 - \exp\left\{-\frac{3 \cdot Wg \cdot E \cdot H \cdot Qw}{4 \cdot r \cdot v_r \cdot Qq}\right\} \tag{20}$$

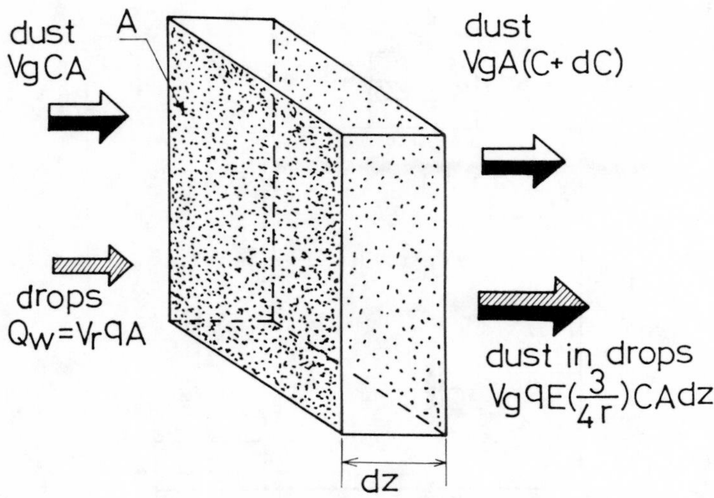

FIGURE 11. Mathematical model for the cocurrent flow.

From this equation we can see that when the upward gas velocity v_g is nearly equal to the sedimentation velocity of the drops, the drop velocity relative to the scrubber is low, then the hold-up is high, and the total collection efficiency η_c becomes high. Therefore, a conservative design for counter-flow would minimize scrubber diameter in order to ensure maximum collection efficiency.

D. Cocurrent Flow

Cocurrent flow occurs in high velocity scrubbers such as the venturi scrubber, as shown in Figure 11. The majority of these involve the atomization of the liquid by the high velocity gas stream. Liquid hold-up and particle collection are described in the countercurrent flow. The particle concentration in a volume element of gas during the process of picking up some water, atomizing it and accelerating the drops to zero relative velocity (i.e., $\xi = 0$) can be described by the following two equations. To avoid the confusion with terminal settling velocity, $V_{gr} = (\xi v_g)$ is used to describe the drop velocity relative to the gas

$$-\frac{dC}{C} = \frac{v_{gr}}{v_g}\cdot\frac{Qw}{(v_g - v_{gr})A}\cdot\frac{3}{4\,r}\cdot E\cdot dz = \frac{3\cdot\xi\cdot E\cdot Qw}{4\cdot(1-\xi)\cdot r\cdot Qg}\cdot dZ \qquad (21)$$

Here, assuming that the quantities ξ and E are independent of Z, Qw, and Qg, we can obtain the following equation

$$\ln\frac{Co}{Ci} = \exp\left\{-\frac{3\cdot\xi\cdot E\cdot Qw\cdot Z}{4\cdot(1-\xi)\cdot r\cdot Qg}\right\} \qquad (22)$$

Therefore, we can obtain the total collection efficiency η_c as

$$\eta_c = \frac{Ci - Co}{Ci} = 1 - \exp\left\{-\frac{3\cdot\xi\cdot E\cdot Qw\cdot Z}{4\cdot(1-\xi)\cdot r\cdot Qg}\right\} \qquad (23)$$

VI. SEPARATION OF AEROSOLS AND HARMFUL GAS BY EJECTOR-SCRUBBER

A. Introduction

Takashima (1961)[14] developed the theoretical characteristics of the ejector-scrubber. In

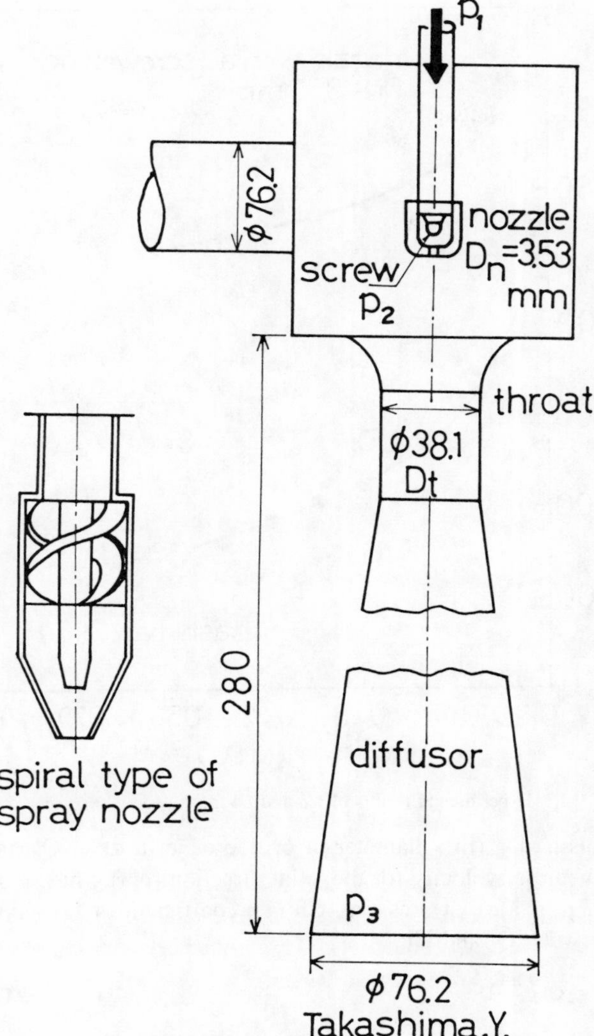

FIGURE 12. Construction of an ejector for the venturi scrubber.

this section, the author wants to describe the theoretical considerations following Professor Takashima's paper.

An ejector for the venturi scrubber has a construction as shown in Figure 12. A nozzle is designed for the nearly homogeneous dispersion of atomization droplets in the throat section. Those droplets help the separation of the harmful gas and also promote the momentum exchange between gas and droplet. Therefore, the jet velocity of droplets, the size distribution of droplets, and the concentration of the droplets are the most important factors for the separation characteristics of the dust and of the harmful gas. The relative velocity between gas velocity and droplet velocity shows the important role. On the other hand, the motion of gas depends on the construction and the size of the throat and the diffusor.

B. Venturi Ejector

One of the most important points is how to make the atomization of liquid by the nozzle. However, it is necessary not only to make the atomization, but also to keep a high jet velocity for the main flow direction.

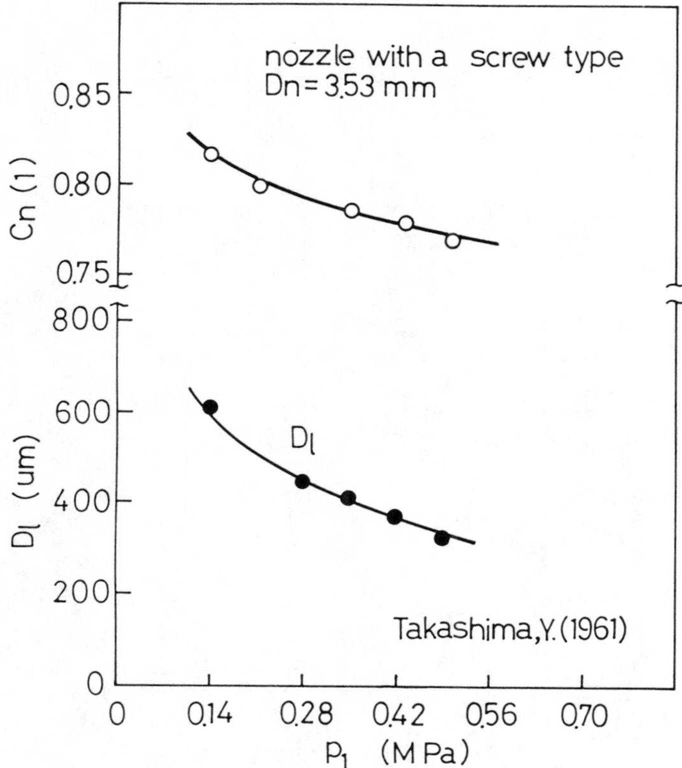

FIGURE 13. Experimental results of Cn and Dl for the screw type of nozzle.

Using the symbols as Dn (m); diameter of orifice of jet nozzle, Ql (m^3/s); flow rate of liquid, Ulo (m/s); mean jet-velocity for the axial direction, ρ_l (kg/m^3); density of liquid, p_l (Pa); feed pressure, p_o (Pa); exit pressure, Cn (1); coefficient of flow rate, we can obtain the relationships between Ql and Ulo:

$$\text{Ulo} = \text{Ql} \Big/ \frac{\pi}{4} \cdot \text{Dn}^2 \tag{24}$$

$$\text{Ql} = \text{Cn} \cdot \frac{\pi}{4} \cdot \text{Dn}^2 \cdot \sqrt{\frac{p_l - p_o}{\rho_l}} \tag{25}$$

where Cn depends on the type of the nozzle. Then the size Dl of the droplets and Cn depend on the pressure difference p_l-p_o. Figure 13 shows the experimental results of Cn and Dl for the nozzle with the screw type by Takashima (1961).

Figure 14 shows the droplet size distributions for the centrifugal spray nozzle by Nelson and Stevens (1961).[15] They described that for all of the materials and nozzle combinations investigated, the data appeared to fit the square-root normal distribution function. This was determined by plotting the cumulative mass vs. the square root of the sieve size on normal probability paper, as shown in Figure 14. Figure 15 shows the size distribution of fuel oil with a hollow cone spray by Glaser (1959). This example of flux data for an angle 80° nozzle and for spraying fuel oil at 12.83 cm^3/s was obtained. Then the velocity Ult of the droplets in the throat is nearly equal to Ulo due to the inertial force of the droplet itself. Therefore, the droplets decrease velocity by contacting the slow velocity of gas. Consequently we can obtain the pressure difference δp by the momentum exchange as follows:

$$\frac{\pi}{4} \cdot D^2 \cdot \delta p = - \text{Ql} \cdot \rho_l \cdot \delta \text{Ul} \tag{26}$$

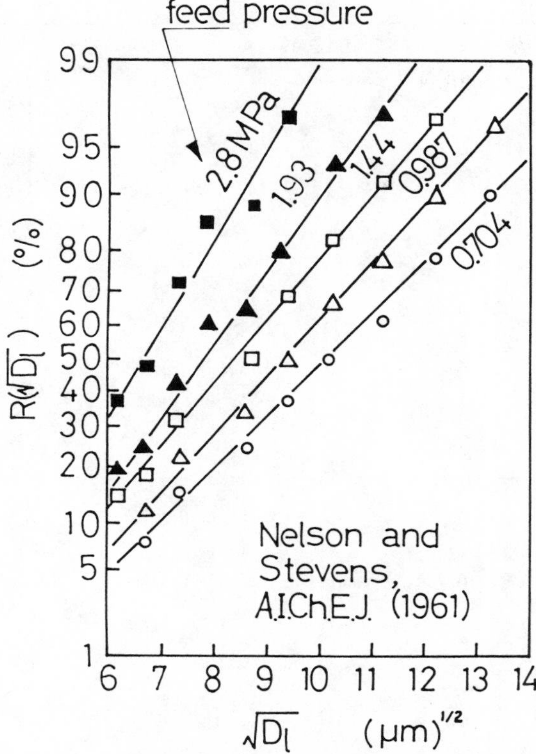

FIGURE 14. Size distributions of the droplets for the centrifugal spray nozzle.

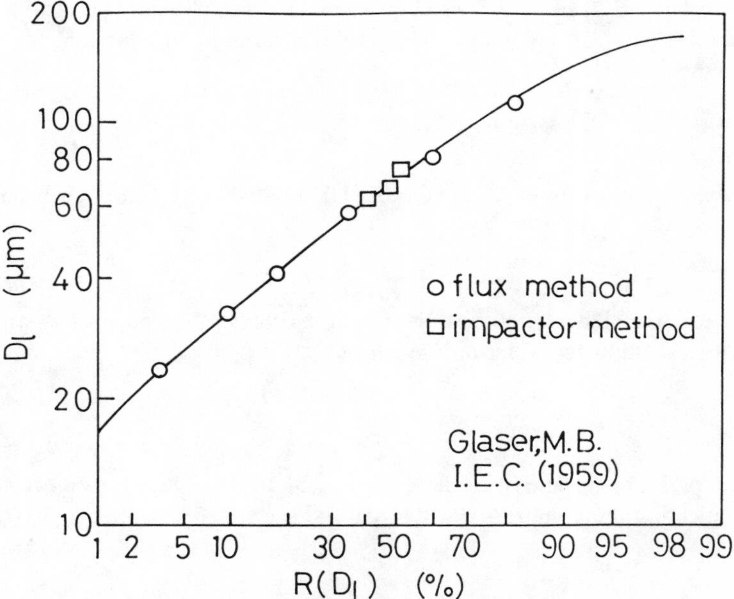

FIGURE 15. Size distribution of fuel oil with hollow cone spray.

where D is the pipe diameter at the arbitrary section. So we can obtain the equation of motion of the droplet which is dependent on the relative velocity Ul − v_g between the droplet velocity Ul and gas velocity v_g as shown in Figure 16, as follows:

$$\frac{dUl}{dt} = -\frac{3}{4} \cdot \frac{C_D}{Dl} \cdot \frac{\rho_g}{\rho_l} \cdot (Ul - v_g)^2 + g \qquad (27)$$

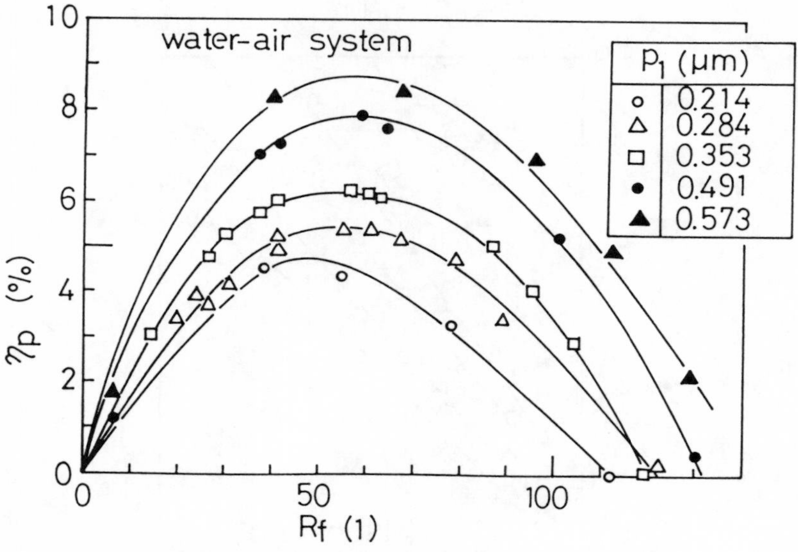

FIGURE 16. Relationship between power efficiency and Rf.

or

$$dUl = \left\{ -\frac{3}{4} \cdot \frac{C_D}{Dl} \cdot \frac{\rho_g}{\rho_l} \cdot (Ul - v_g)^2 + g \right\} \cdot \frac{dX}{Ul} \tag{28}$$

where C_D is the drag coefficient for the droplet which is related to Reynolds number of droplet diameter Dl, and X is the downstream distance based upon the point of the entrance place at the throat. Now, in order to solve the above differential equation, we must use a mass balance equation

$$Ql + Qg = \frac{\pi}{4} \cdot D^2 \cdot (Ul + v_g) \tag{29}$$

From the practical point of view, $Qg \gg Ql$ and $Ul \gg v_g$, then Equation 29 becomes

$$v_g \doteqdot Qg \Big/ \frac{\pi}{4} \cdot D^2 \tag{30}$$

On the other hand, the characteristics of the pressure drop and the power loss are described by the following equation for water-air ejector as

$$Rf = \frac{Qg}{Ql}, \quad Rp = \frac{p_3 - p_2}{p_1 - p_3} \tag{31}$$

where p_3 is the pressure at the exit of the diffusor and also the power efficiency η_p may be defined for neglecting the compressibility of gas as

$$\eta_p = Rf \cdot Rp \tag{32}$$

Takashima showed the experimental results of power efficiency η_p for the water-air system for the various kinds of feed pressures p_1, as shown in Figure 17. From this figure we will find that the maximum power efficiency η_p is obtained at Rf $\doteqdot$ 50 which is independent of the feed pressure p_1.

Then Takashima showed the theoretical consideration of driving condition of how to

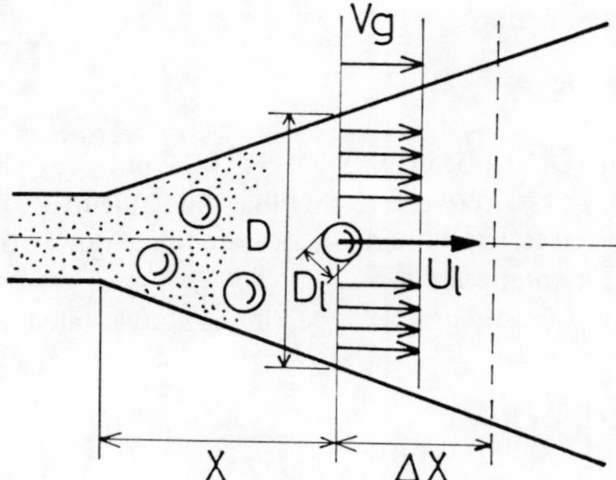

FIGURE 17. Separation mechanism of the particles by the droplets.

obtain the maximum power efficiency. Combining Equations 26, 27, and 30, we can obtain the following equation

$$\frac{dp}{dX} = \frac{Ql \cdot \rho_l}{\pi \cdot D^2/4} \cdot \frac{dUl}{dX} \doteqdot \frac{3}{4} \cdot \frac{\rho_g}{Rf \cdot Dl} \cdot C_D \cdot \frac{v_g}{Ul} \cdot (Ul - v_g)^2 \tag{33}$$

Here we can neglect the effect of gravity for the motion of droplets. From Equation 33, when the quantities Rf and Dl are given, the drag coefficient C_D becomes $C_D = 24/Rep$ for $Rep = Dl \cdot (Ul - v_g)\, \rho_g/\eta \leqq 4$, $C_D = 13/\sqrt{Rep}$ for $10 \leqq Rep \leqq 1000$ and $C_D \doteqdot 0.4$ for $Rep \geqq 1000$. Accordingly, we can obtain the conditions from Equation 33 for Rf to be maximum as follows:

$$\left.\begin{array}{ll} \text{in the case of} \quad Rep \leqq 4, & v_g = \dfrac{1}{2} \cdot Ul \\ \text{in the case of} \quad 10 \leqq Rep \leqq 1000, & v_g = \dfrac{2}{5} \cdot Ul \\ \text{in the case of} \quad Rep \geqq 1000, & v_g = \dfrac{1}{3} \cdot Ul \end{array}\right\} \tag{34}$$

Now, designing the throat (Dt) and diffusor portions based upon satisfying the conditions of Equation 34, the size of the throat and diffusor for the most effective pressure recovery per unit length can be attained as follows:

in the case of $Rep \leqq 4$,

$$1 - \left(\frac{Dt}{D}\right)^2 = \frac{9}{Reo} \cdot \frac{\rho_g}{\rho_l} \cdot \frac{X}{Dl}, \quad Ra = \left(\frac{Dt}{Dn}\right)^2 = 2\ Rf \tag{35}$$

in the case of $10 \leqslant Rep \leqslant 1000$,

$$1 - \frac{Dt}{D} = \frac{2.27}{\sqrt{Reo}} \cdot \frac{\rho_g}{\rho_l} \cdot \frac{X}{Dl}, \quad Ra = \left(\frac{Dt}{Dn}\right)^2 = 2.5\ Rf \tag{36}$$

in the case of $Rep \geqq 1000$,

$$\frac{Dt}{D} = \exp(-0.075) \frac{\rho_g}{\rho_l} \cdot \frac{X}{Dl}, \quad Ra = \left(\frac{Dt}{Dn}\right)^2 = 3.0\ Rf \tag{37}$$

where a symbol Reo is defined as

$$\text{Reo} = \frac{\text{Dl} \cdot \text{Ulo} \cdot \rho_g}{\eta} \tag{38}$$

Thereafter in the ejector (Dn = 3.53 mm, Dt = 38.1 mm), as shown in Figure 12, the value of Ra becomes Ra = 116. We can expect that the value of Rf for the maximum power efficiency η_p exists between Rf = 40 and 60, as shown in Figure 16. This theoretical result coincides with the experimental results.

The power efficiency η_p of an ejector designed in the above stated manner can be expressed as

$$\eta_p = \frac{\text{Cn}^2}{\text{k}} \left\{ 1 - \left(\frac{\text{Dt}}{\text{D}} \right)^4 \right\} \tag{39}$$

where the values of k are 2.0, 2.5, and 3.0 corresponding to the values of Rep.

C. Separation Mechanism of Particles

We consider the probability of one particle escaping from this scrubber without droplet collisions, as shown in Figure 17. Assuming that the size of the particle is equal or less than Xp ≒ 1 μm, so the motion of this particle may follow the motion of fluid flow. Therefore, the velocity of the particle Up is equal to gas velocity v_g. Assuming that number of droplets per unit volume at the arbitrary distance X for the downstream direction from the throat position is n, we can obtain a relationship between Ql and n

$$n = \frac{\text{Ql}}{\frac{\pi}{6} \text{Dl}^3 \cdot \text{Ul} \cdot \frac{\pi}{4} \text{D}^2} \tag{40}$$

Therefore, while the particle moves to v_g, the particle will be passed by droplets in numbers as

$$n(\text{Ul} - v_g) \cdot \frac{\pi}{4} \text{D}^2 \tag{41}$$

The total cross-sectional area of these droplets per unit area ($\pi D^2/4$) can be written

$$\frac{n(\text{Ul} - v_g) \cdot \frac{\pi}{4} \text{D}^2 \cdot \frac{\pi}{4} \text{Dl}^2}{\frac{\pi}{4} \text{D}^2} = \frac{3}{2} \cdot \frac{\text{Ql}}{\text{Qg}} \cdot \frac{1}{\text{Dl}} \cdot \frac{v_g}{\text{Ul}} \cdot (\text{Ul} - v_g) =$$

$$= \frac{3}{2} \cdot \frac{1}{\text{Rf}} \cdot \frac{\text{Ul} - v_g}{\text{Ul}} \cdot \frac{v_g}{\text{Dl}} \tag{42}$$

Consequently, the probability of particle collision with the droplets during the particle movement $\Delta X = v_g \cdot \Delta t$ can be expressed as

$$\frac{3}{2} \cdot \frac{1}{\text{Rf}} \cdot \frac{\text{Ul} - v_g}{\text{Ul}} \cdot \frac{\Delta X}{\text{Dl}} \cdot \eta_t \tag{43}$$

where η_t means target efficiency. In consequence, the probability of avoiding the collision of the particle with the droplets becomes

$$\lim_{\Delta X \to 0} \left(1 - \frac{3}{2} \cdot \frac{1}{\text{Rf}} \cdot \frac{\text{Ul} - v_g}{\text{Ul}} \cdot \frac{\Delta X}{\text{Dl}} \cdot \eta_t \right)^{X/\Delta X} =$$

$$= \exp\left(-\frac{3}{2} \cdot \frac{1}{\text{Rf}} \cdot \frac{\text{Uf} - v_g}{\text{Ul}} \cdot \frac{X}{\text{D1}} \cdot \eta_t \right) \tag{44}$$

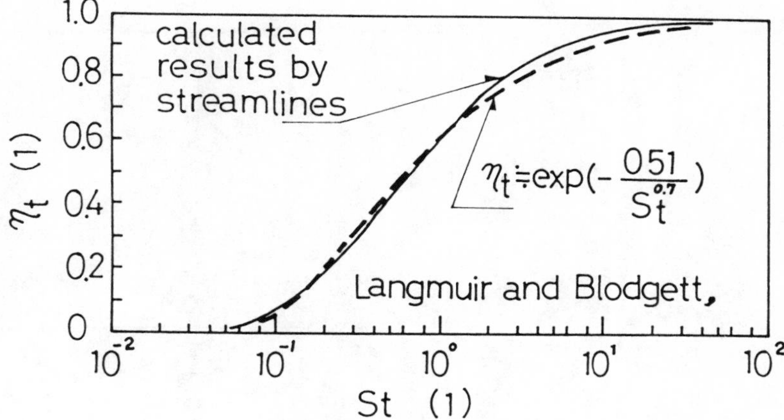

FIGURE 18. Target efficiency for a sphere.

Then the collection efficiency η_c of the ejector-scrubber can be expressed as

$$\eta_c = 1 - \exp\left(-\int_0^X \frac{3}{2}\cdot\frac{1}{Rf}\cdot\frac{Ul - v_g}{Ul}\cdot\eta_t\cdot\frac{\delta X}{D1}\right) \tag{45}$$

So we may find from Equation 45 that η_c is increased with increasing the velocity difference ($Ul - v_g$). Also from the designing point of view, a condition of the maximum power efficiency Ra = 2 to 3 Rf is best for the particle separation. Now the target efficiency η_t, as shown in Figure 18, can be estimated by the equation of Langmuir and Blodgett (U.S.A.A.F.T.R., 5418, (1946)) as

$$\eta_t \doteqdot \exp\left(-\frac{0.51}{St}\right) \tag{46}$$

where St is Stokes number defined as

$$St = \frac{(Ul - v_g)\,Wsg}{Dl\cdot g} \tag{47}$$

In this equation Wsg means the terminal velocity of the particle in gas defined as

$$Wsg = \frac{\rho_p\cdot Xp^2\cdot g}{18\cdot\eta}\cdot\left(1 + Km\,\frac{\lambda m}{Xp}\right) \tag{48}$$

where Km is Cunningham's correction factor and λm is mean free path of gas molecules.

D. Numerical Example

Using an ejector as shown in Figure 12, we try to calculate the target efficiency η_t and the total collection efficiency η_c for the particle floating in air.

Assuming that the particle sizes are Xp = 0.25, 0.5, and 1 μm and the particle density is $\rho_p = 2$ g/cm^3.

For the solution we obtain $(Ul - v_g)/Ul = 0.6$ for $Ul = 2.5\,v_g$. Then we can obtain Equation 45 for Rf = 45 and Dl = 200 μm as follows

$$\eta_c = 1 - \exp\left(-\frac{3\times 0.6}{2\times 45\times 0.02}\cdot\int_0^X \eta_t\cdot\delta X\right) = 1 - \exp\left(-\int_0^X \eta_t\cdot dX\right)$$

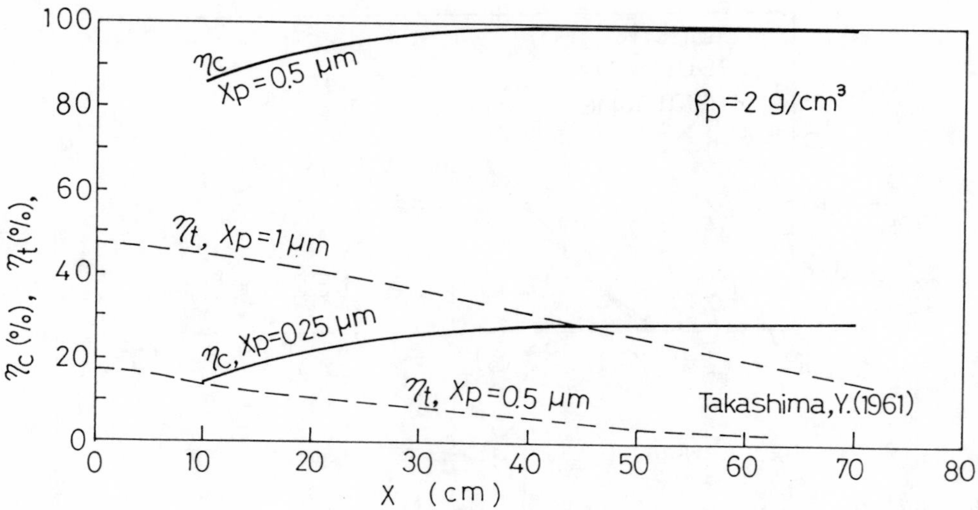

FIGURE 19. Relationship between η_t, η_c, and X.

On the other hand, from Equation 35, we can obtain the following equations

$$1 - \left(\frac{v_g}{v_{gt}}\right)^{0.5} = 0.007 \cdot X$$

$$v_{gt} = \frac{Ulo}{2.5} = \frac{2800}{2.5} = 1120$$

Finally, we can get the following relation

$$Ul - v_g = 1680 \times (1 - 0.007 \cdot X)^2$$

Therefore, Stokes number St becomes

$$St = \frac{1680 \times (1 - 0.007 \cdot X)^2 \cdot Wsg}{980 \times 0.02} = 85.7 \times (1 - 0.007 \cdot X)^2 \cdot Wsg$$

The target efficiency η_t becomes as follows

$$\eta_t = \exp\left(-\frac{0.51}{410.7}\right) = \exp\left\{-\frac{0.0225}{(1 - 0.007 \cdot X)^{1.4} \cdot Wsg^{0.7}}\right\}$$

Now the numerical results of Wsg for the normal pressure and normal temperature can be written as follows:

Xp (μm)	1	0.5	0.25
Wsg(cm/s)	6.9×10^{-3}	1.97×10^{-3}	6.24×10^{-4}

Consequently the relationships between η_t, η_c, and X are shown in Figure 19. From Figure 19, we will find that η_c becomes 99.5% for Xp = 0.5 μm but η_c becomes 30% for Xp = 0.25 μm, and still more, η_c becomes nearly 0% for Xp = 0.25 μm. In this ejector the value of η_c will be determined within $X \leqq 50$ cm.

VII. SPRAY TOWER WITH GLASS SPHERE BED

Wicke and Krebs[4] investigated the separation performance and pressure drop for the spray

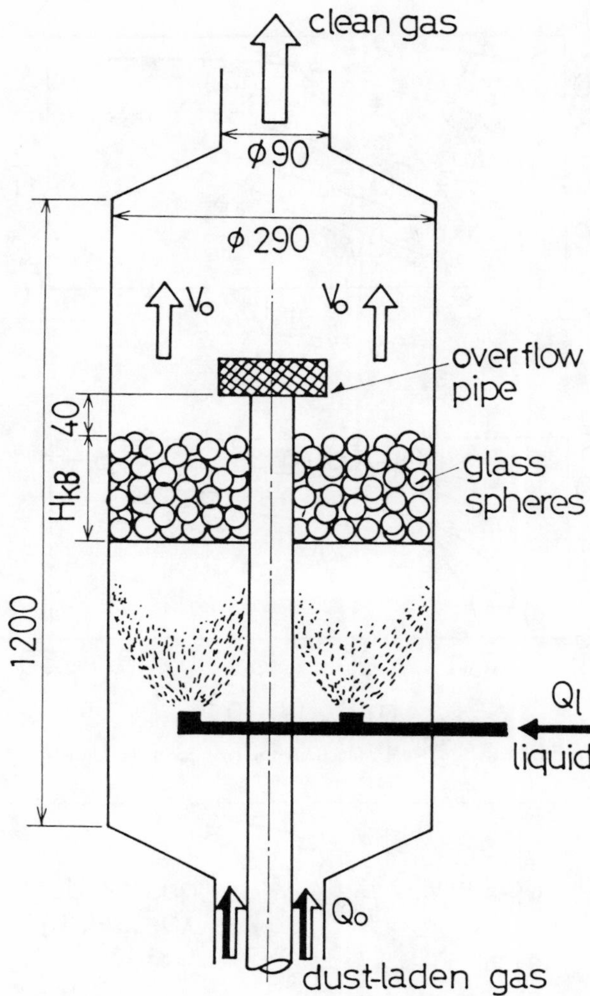

FIGURE 20. Spray tower with glass-sphere bed.

tower with glass sphere (10 mm) bed, as shown in Figure 20. They designed the upward velocity Vo to be about up to 2.5 m/s for vibrating the glass spheres. And, at the same time the liquid droplets with dust-laden gas flow up through the glass sphere bed.

Figure 21 shows the total collection efficiency η_c and the pressure drop Δp_c for the various glass sphere bed height Hk*B* and liquid flow rate Ql for the quartz dust (Xp = 2.7 μm). Figure 22 shows the fractional collection efficiencies η_x (Xp) for Hk*B* = 200, 70, and 30 mm under the test conditions of Ql = 200 1/h, dust-concentration Co = 1 g/cm^3, Hs*B* = 40 mm, and the upward gas velocity Vo = 1.13 m/s.

From this figure, we will find that the collection efficiency η_c strongly depends on the height Hk*B* of the glass sphere bed and also the cut-size Xc is between Xc = 1.13 and 1.3 μm and that the particle size Xp below 0.5 μm cannot be separated.

VIII. MULTI-STAGE COMPACT SCRUBBER WITH CONTROLLABLE HYDRAULIC RESISTANCE

A new and patented type of a two-stage collision wet scrubber[16] for high pressures, which is equipped with the internal fluid circulation and a wide range control of the hydraulic

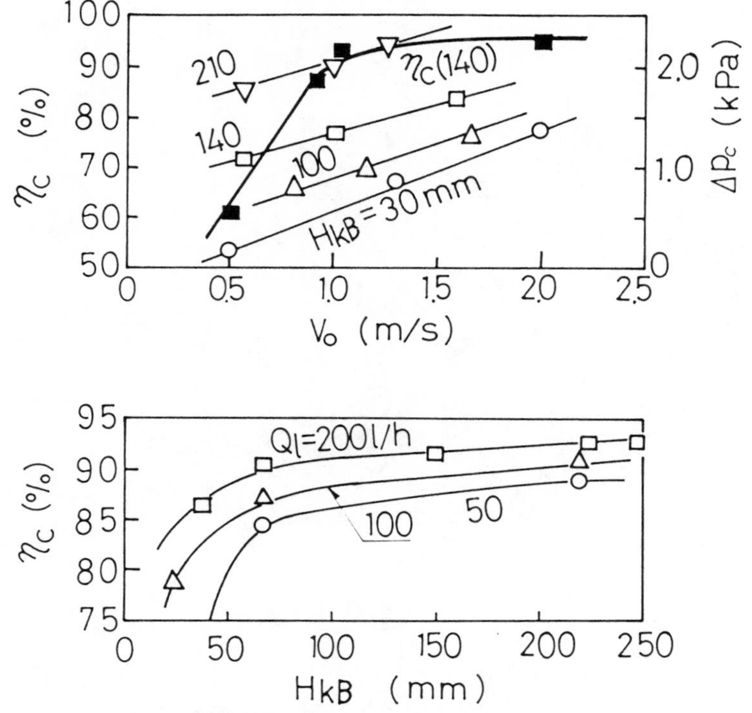

FIGURE 21. Total collection efficiency for quartz dust.

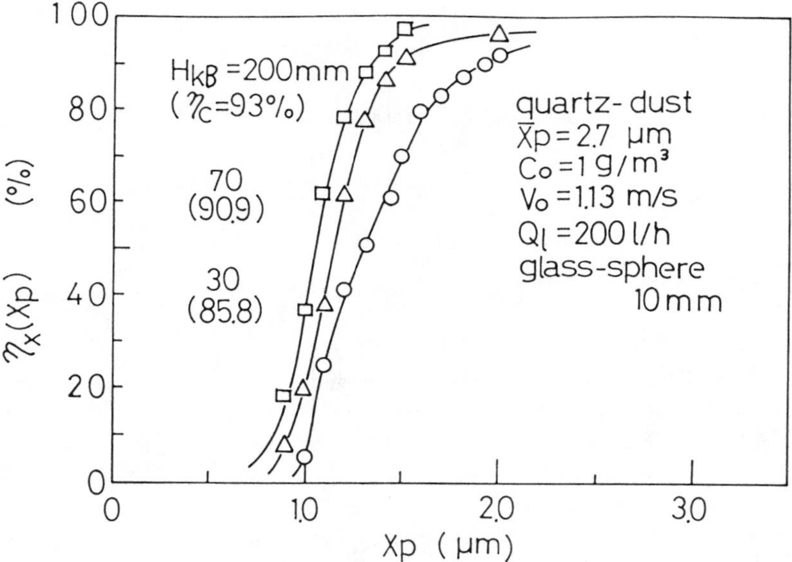

FIGURE 22. Experimental result of the fractional collection efficiency.

resistance, is shown in Figure 23. This type of collector has high collection efficiency η_c for the particle size Xp larger than 2 μm.

The dust-laden gas flows through the first and second regions of collision with water droplets. The water droplets which are brought with the upward flow air are separated at the droplet separator. Then the total pressure drop Δp_c is expected between 4000 to 8000 Pa for flow rate Qo = 8500 to 3500 m³/hr.

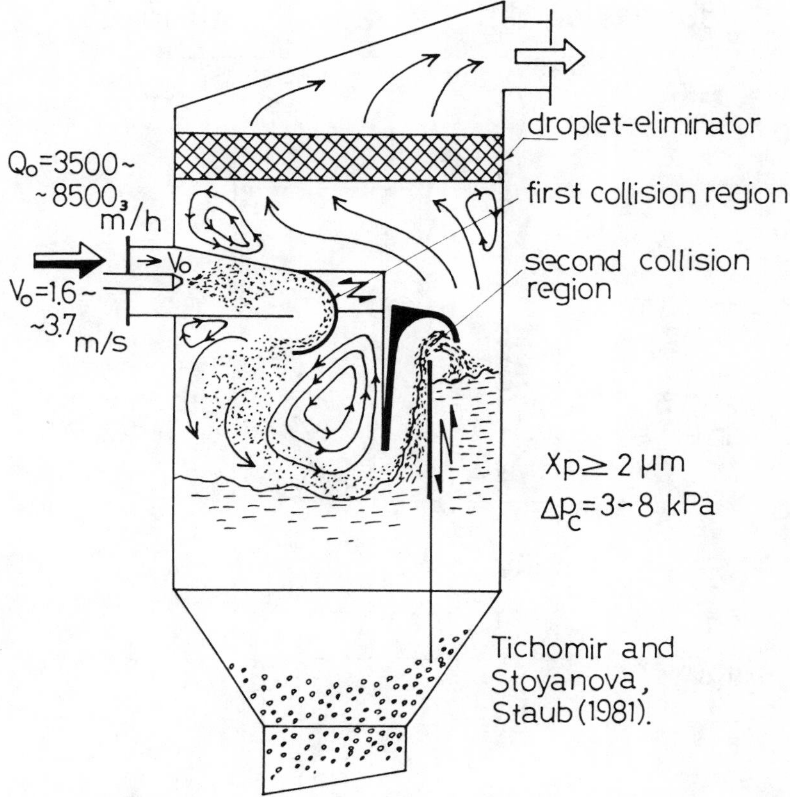

FIGURE 23. Multi-stage compact scrubber.

IX. VENTURI-LOUVER SEPARATOR

Štorch[17] (1966) developed a new type of venturi-louver separator for separating dust of a particle size smaller than Xp = 1 μm at Forschungsinstitut für Lufttechnik in Prague, as shown in Figure 24. The collection efficiency η_c did not vary for water supply between 2 ℓ/m^3 and 5 ℓ/m^3.

Concerning the collection efficiency on the concentration of the dust-laden gas Co = 100 to 150 mg/m^3 for the test dust with 60% under 1 μm and 25% under 0.5 μm in the research room, η_c could reach 80% on the pressure drop $\Delta p_c = 3.43$ kPa of venturi-pipe, $\eta_c = 93\%$ on $\Delta p_c = 2.94$ kPa and $\eta_c = 96\%$ on $\Delta p_c = 6.86$ kPa, respectively.

Figure 25 shows the experimental results of the collection efficiency η_c for test dust of 90% FeSi and for dust-concentration Co = 0.5 to 4 g/m^3 in the case of the gas temperature 423 K.

X. DEPRESSION OF LIQUID SURFACES BY GAS JETS

Collins and Lubanska[18] (1954) investigated the maximum depth P(cm) of penetration of the liquid surface by a gas jet directed at the liquid surface, as shown in Figure 26. They derived an equation of penetration P(cm) as a function of thrust and an angle of jet as follows:

$$P = \frac{53\ T \sin\theta}{\left\{X^2 \cdot \rho_l \cdot g + 19(\rho_l \cdot g \cdot T^2)^{1/3}\right\}} \tag{49}$$

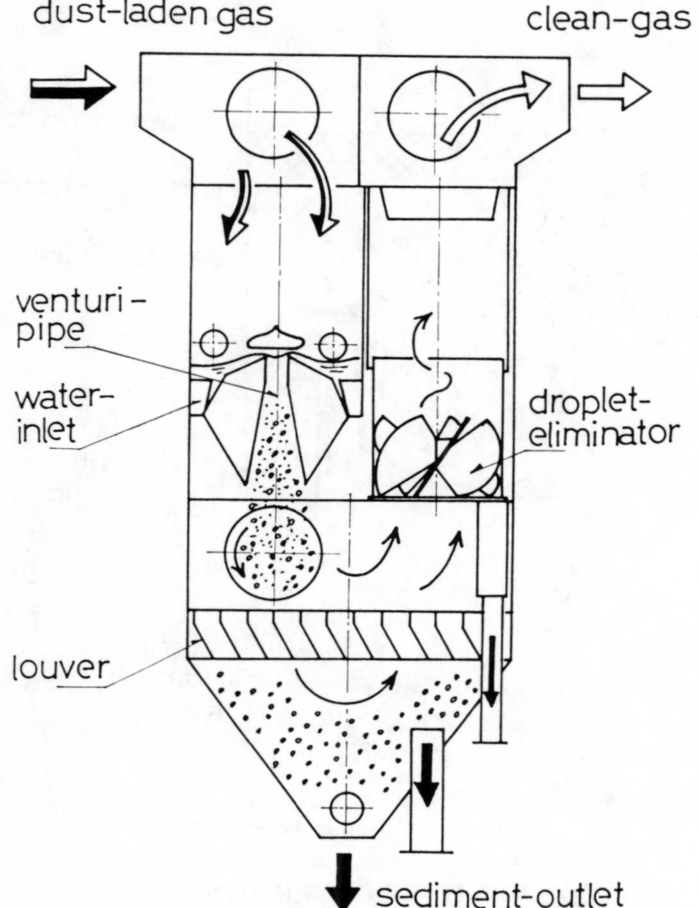

FIGURE 24. Venturi-louver separator.

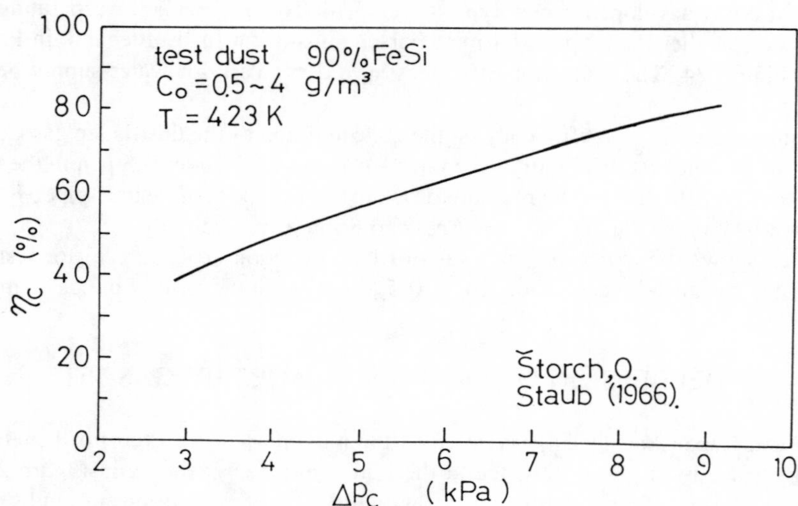

FIGURE 25. Experimental result of the collection efficiency.

where P (cm) is a maximum depth of depression, X (cm) is slant distance of jet orifice from undisturbed water surface, θ (deg) is an angle between jet axis and projection on water

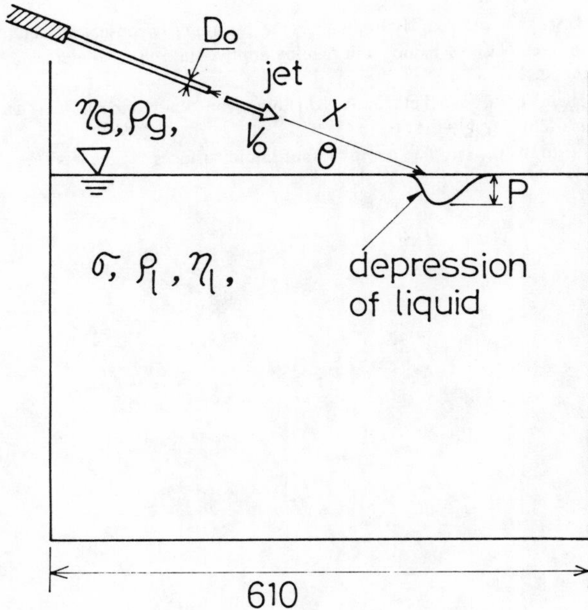

FIGURE 26. Penetration of the liquid surface by a gas jet.

surface, T (dynes) is reaction thrust ($\pi \cdot \rho_g \cdot Do^2 \cdot Vo^2/4$), ρ_g (g/cm^3) is density of gas, ρ_l (g/cm^3) is density of liquid, Vo (m/s) is mean velocity of gas at the jet mouth, σ (dynes/cm) is the surface tension of liquid, and Do (cm) is a diameter of jet orifice.

REFERENCES

1. **Bethea, R. M.,** *Air Pollution Control Technology,* Van Nostrand, New York, 1978.
2. **Ooyama, Y.,** *Chemical Engineering II* (in Japanese), Iwanami Publishing, Tokyo, 1963.
3. **Kleinschmidt, R. V.,** in *Chem. Metall. Eng.,* 46, 487, 1939.
4. **Wicke, M. and Krebs, F. E.,** Vergleichende Versuche Zum Betriebsverhalten und zur Abscheideleistung von Naβentstaubern, *Chem. Ing. Tech.,* 43(6), 386, 1971.
5. **Kleinschmidt, R. V. and Anthony, A. W.,** Recent developments of the Pease-Anthony gas scrubber, *Trans. Am. Soc. Mech. Eng.,* May, 349, 1941.
6. **Engle, M. D.,** Pease-Anthony gas scrubbers, *Trans. Am. Soc. Mech. Eng.,* 59(10), 358, 1937.
7. **Hinze, J. O.,** Fundamentals of the hydrodynamic mechanism of splitting in dispersion processes, *Am. Inst. Chem. Eng. J.,* 1(3), 289, 1955.
8. **Clay, P. H.,** in *Proc. R. Acad. Sci. (Amsterdam),* 43, 852, 1940.
9. **Goldschmidt, Y. and Calvert, S.,** Small particle collection by supported liquid drops, *Am. Inst. Chem. Eng. J.,* 9(3), 352, 1963.
10. **Walton, W. H. and Woolcock, A.,** in *Inst. J. Air Pollut.,* 3, 129, 1960.
11. **Fonda, A. and Herne, H.,** *Inst. J. Air Pollut.,* 3, 26, 1960.
12. **Stern, A. C.,** *Air Pollution,* Vol. 3, 2nd ed., Academic Press, New York, 1968, 457.
13. **Crawford, M.,** *Air Pollution Control Theory,* McGraw-Hill, New York, 1976.
14. **Takashima, Y.,** Separation of aerosols and harmful gas by ejector-scrubber, *Saikin-no Kagaku-Kogaku* (in Japanese), Maruzen Publishing, Tokyo, 1961.
15. **Nelson, P. A. and Stevens, W. F.,** Size distribution of droplets from centrifugal spray nozzles, *Am. Inst. Chem. Eng. J.,* 7(1), 80, 1961.

16. **Tichormir, S. and Stoyanova, A.,** Hydrodynamische Strömungswechselwirkungen in den Kontaktelementen eines mehrstufigen Naβabscheiders mit regelbarem hydraulischem Widerstand, *Staub,* 41(11), 433, 1981.
17. **Štorch, O.,** Ein neuer Venturi-Wächer zur Abscheidung von Staubteilchen < 1 μm, insbesondere von brauern Rauch, *Staub,* 26(11), 479, 1966.
18. **Collins, R. D. and Lubanska, H.,** The depression of liquid surfaces by gas jets, *Br. J. Appl. Phys.,* 5(1), 22, 1954.

APPENDIX OF MAIN SYMBOLS

CHAPTER ONE

Ab	(1)	Separation index
Ao	(m^2)	Cross-sectional area of the inlet pipe
a	(m)	Radius corresponding to the maximum tangential velocity
Co	(kg/m^3)	Feed dust concentration
D1	(m)	Cyclone diameter
G	(N)	Gravity force
l	(m)	Mixing length
m	(1)	Separation index
n	(1)	Velocity index of the quasi-free vortex
p	(Pa)	Static pressure
Δp_c	(Pa)	Pressure drop
Qo	(m^3/s)	Flow rate of gas
Reb	(1)	Reynolds number defined by Equation 8
Rcy	(1)	Cyclone Reynolds number defined by Equation 11
Uθ	(m/s)	Tangential velocity of the solid particle
Vo	(m/s)	Mean inlet velocity of gas
Vθ	(m/s)	Tangential velocity of gas or air
Xc	(μm)	Cut-size
Xp	(μm)	Particle diameter
Wsc	(m/s)	Centrifugal sedimentation velocity
Z	(N)	Centrifugal force
Φ	(1)	Inertia parameter
Γ	(m^2/s)	Circulation
ω	(rad/s)	Angular velocity
ρ	(kg/m^3)	Density of gas
ρ_p	(kg/m^3)	Density of the solid particle
ζ_c	(1)	Coefficient of the pressure drop
υ_t	(m^2/s)	Eddy kinematic viscosity
η_c	(%)	Total collection efficiency
η_x	(1)	Fractional collection efficiency
τ_t	(Pa)	Reynolds stress

CHAPTER TWO

Co	(g/m^3)	Feed dust concentration
D1	(m)	Outer diameter of the separation chamber
D2	(m)	Diameter of the inner exit pipe
Ds	(m)	Diameter of the slit
Dv1	(m)	Diameter of the exit nozzle of the primary vortex chamber
H	(m)	Height from the primary vortex chamber to the inner exit pipe
m	(1)	Separation index
Δp_1	(Pa)	Pressure drop of the primary gas flow
Δp_2	(Pa)	Pressure drop of the secondary gas flow
Δp_c	(Pa)	Total pressure drop
Q1	(m^3/s)	Flow rate of the primary gas flow

Q2	(m^3/s)	Flow rate of the secondary gas flow
R(Xp)	(1)	Cumulative distribution of particles
Rmi	(m)	Representative radius at the inlet pipe
Vθ	(m/s)	Tangential velocity of gas flow
Vz	(m/s)	Axial velocity of gas flow
Vi	(m/s)	Mean inlet velocity in the inlet pipe
Xp	(μm)	Particle size
Xc	(μm)	Theoretically calculated cut-size
Xc50	(μm)	Experimentally determined cut-size
X_{R50}	(μm)	Representative diameter of the dust corresponding to the particle size R(Xp) = 50%
η	(Pa·s)	Viscosity of gas
η_c	(%)	Total collection efficiency
η_x	(1)	Fractional collection efficiency
ρ_p	(kg/m^3)	Density of the particle
ρ	(kg/m^3)	Density of gas
Ψ	(1)	Dimensionless equi-flow rate line

CHAPTER THREE

C_D	(1)	Drag coefficient
C	(1)	Packing density
D	(m)	Target diameter
L	(m)	Length of all of the fibers
m	(g/m^2)	Dust load
n	(particles/m^3)	Particle numbers per unit volume
P	(1)	Penetration
Po	(1)	Porosity of the fibrous filter
Δp	(Pa)	Pressure drop
Qo	(m^3/s)	Flow rate of gas
Rep	(1)	Flow Reynolds number around a fiber
R	(μm)	Fiber radius
St	(1)	Stokes number
$\overline{Up}$	(cm/s)	Migration velocity in the electric field
$\overline{V}$	(cm/s)	Filtration velocity or mean gas velocity
Xp	(μm)	Particle diameter
γ	(1/m)	Filtration index
ϵ	(1)	Void
η	(Pa·s)	Viscosity of gas
η_c	(%)	Collection efficiency
ρ_p	(kg/m^3)	Particle density
Ψ	(1)	Stokes number
Ω	(m/s^2)	Vorticity

CHAPTER FOUR

Cm	(1)	Cunningham's correction factor
Co	(kg/m^3)	Dust concentration
D2	(m)	Diameter of the discharge electrode
E	(kV/cm)	Intensity of the electric field

Symbol	Unit	Meaning
Fe	(N)	Coulomb force
I	(A/cm)	Discharge current per unit length of the discharge electrode
Kp	(cm/s/V/cm)	Mobility of the solid particle
m*p*	(kg)	Mass of the solid particle
n	(1)	Number of electron charge
p	(Pa)	Pressure of gas
q	(C)	Electric charge
Rc	(cm)	Range of the corona sheath
T	(K)	Gas temperature
$\overline{Up}$	(m/s)	Migration velocity of the charged solid particle
Qo	(m/s)	Flow rate
$\overline{Vo}$	(m/s)	Mean gas velocity
Wsg	(cm/s)	Terminal velocity of the solid particle
Xp	(μm)	Diameter of the solid particle
ρ	(Ω-cm)	Resistivity
λ	(cm)	Mean free path of the gas
η	(Pa·s)	Viscosity of gas
η_c	(%)	Total collection efficiency

CHAPTER FIVE

Symbol	Unit	Meaning
C	(g/m^3)	Dust concentration
C_D	(1)	Drag coefficient
Cn	(1)	Cunningham slip correction factor
D	(m)	Pipe diameter
Dp	(μm)	Diameter of globule
P	(m)	Depression of liquid
p	(Pa)	Static pressure
Δp_c	(Pa)	Pressure drop
Ql	(m^3/s)	Flow rate of liquid
Qg	(m^3/s)	Flow rate of gas
St	(1)	Stokes number
Xp	(μm)	Particle diameter
X	(m)	Distance
η_c	(%)	Total collection efficiency
η	(Pa·s)	Viscosity of globule or power efficiency
η_t	(%)	Target efficiency
ρ_p	(kg/m^3)	Density of particle or density of globule
σ	(N/m)	Interfacial tension
τ	(N/m^2)	Surface force per unit area

INDEX

A

B

C

D

E

F

G

H

I

J

K

L

S

T

V

W